Edexcel A2 Physics Revision Guide

for SHAP and concept-led approaches

REVISION GUIDE

A PEARSON COMPANY

Published by Pearson Education Limited, a company incorporated in England and Wales, having its registered office at Edinburgh Gate, Harlow, Essex, CM20 2JE. Registered company number: 872828

Edexcel is a registered trade mark of Edexcel Limited

First published 2009

12 11 10 09
10 9 8 7 6 5 4 3 2 1

British Library Cataloguing in Publication Data
A catalogue record for this book is available from the British Library

ISBN 978 1 846905 94 0

External project management by Gillian Lindsey
Edited by Geoff Amor
Typeset by Pantek Arts Ltd
Illustrated by Pantek Arts Ltd, Maidstone, Kent
Cover photo © Shutterstock: William Attard McCarthy

Printed in Great Britain by Henry Ling Ltd, at the Dorset Press, Dorchester, Dorset

Acknowledgements
Where exam questions are taken from papers specified at the end of the question, these are reproduced by kind permission of Edexcel.

The publishers are grateful to Damian Riddle for writing the Answering multiple choice and extended questions, Anne Scott and Elizabeth Swinbank at University of York Science Education Group for writing the Revision techniques section and to Tim Tuggey and Andrea Gostick for their collaboration in reviewing this book.

Every effort has been made to contact copyright holders of material reproduced in this book. Any omissions will be rectified in subsequent printings if notice is given to the publishers.

Contents

How to use this Revision Guide

Welcome to your **Edexcel A2 Physics Revision Guide**, perfect whether you're studying Salters Horners Advanced Physics (the blue book), or the 'concept-led' approach to Edexcel Physics (the red book).

This unique guide provides you with tailored support, written by Senior Examiners. They draw on real 'ResultsPlus' exam data from past A-level exams, and have used this to identify common pitfalls that have caught out other students, and areas on which to focus your revision. As you work your way through the topics, look out for the following features throughout the text:

ResultsPlus Examiner Tip
These sections help you perform to your best in the exams by highlighting key terms and information, analysing the questions you may be asked, and showing how to approach answering them. All of this is based on data from real-life A-level students!

ResultsPlus Watch Out!
The examiners have looked back at data from previous exams to find the common pitfalls and mistakes made by students – and guide you on how to avoid repeating them in *your* exam.

Quick Questions
Use these questions as a quick recap to test your knowledge as you progress.

Thinking Task
These sections provide further research or analysis tasks to develop your understanding and help you revise.

Worked Examples
The examiners provide step-by-step guidance on complex equations and concepts.

Each section also ends with:

Section Checklist
This summarises what you should know for this section, which specification point each checkpoint covers and where in the guide you can revise it. Use it to record your progress as you revise.

ResultsPlus Build Better Answers
Here you will find sample exam questions with exemplar answers, examiner tips and a commentary comparing both a basic and an excellent response: so you can see how to get the highest marks.

Practice exam questions
Exam-style questions, including multiple-choice, offer plenty of practice ahead of the written exams.

Both Unit 4 and Unit 5 conclude with a **Practice unit test** to test your learning. These are not intended as timed, full-length papers, but provide a range of exam-style practice questions covering the range of content likely to be encountered within the unit test.

The final Unit consists of advice and support on experimental physics skills, giving guidance on your assessed practical work to help you plan, implement and analyse your own experiment.

Answers to all the in-text questions, as well as detailed, mark-by-mark answers to the practice exam questions, can be found at the back of the book.

We hope you find this guide invaluable. Best of luck!

Revision techniques

Getting started can be the hardest part of revision, but don't leave it too late. Revise little and often! Don't spend too long on any one section, but revisit it several times, and if there is something you don't understand, ask your teacher for help.

Just reading through your notes is not enough. Take an active approach using some of the revision techniques suggested below.

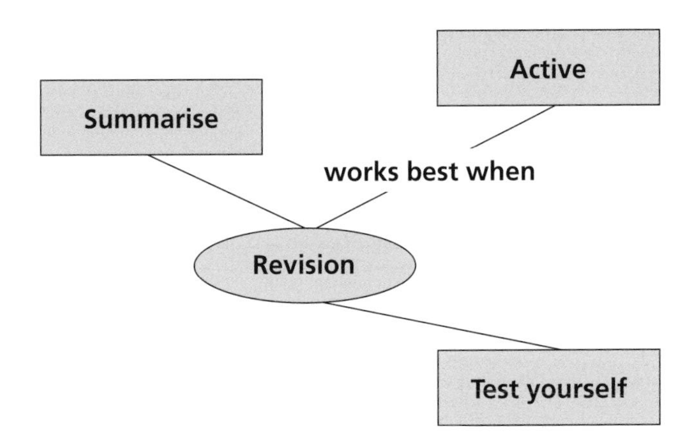

Summarising key ideas

Make sure you don't end up just copying out your notes in full. Use some of these techniques to produce condensed notes.

- Tables and lists to present information concisely.
- Index cards to record the most important points for each section.
- Flow charts to identify steps in a process.
- Diagrams to present information visually.
- Spider diagrams and concept maps to show the links between ideas.
- Mnemonics to help you remember lists.
- Glossaries to make sure you know clear definitions of key terms.

Include page references to your notes or textbook. Use colour and highlighting to pick out key terms.

Active techniques

Using a variety of approaches will prevent your revision becoming boring and will make more of the ideas stick. Here are some methods to try.

- Explain ideas to a partner and ask each other questions.
- Make a podcast and play it back to yourself.
- Use PowerPoint® to make interactive notes and tests.
- Search the Internet for animations, tests and tutorials that you can use.
- Work in a group to create and use games and quizzes.

If you use resources from elsewhere, make sure they cover the right content at the right level.

Test yourself

Once you have revised a topic, you need to check that you can remember and apply what you have learnt.

- Use the questions from your textbook and this revision guide.
- Get someone to test you on key points.
- Try some past exam questions.

Answering multiple-choice and extended questions

Section A of Unit tests 4 and 5 contain objective test (multiple-choice) questions. Section B contains a mixture of short-answer and extended-answer questions, including the analysis, interpretation and evaluation of experimental and investigative activities. In both sections you may be required to apply your knowledge and understanding of physics to situations that you have not seen before.

Multiple-choice questions

For each question there are four possible answers, labelled A, B, C and D. A good multiple-choice question (from an examiner's point of view) gives the correct answer and three other possible answers, which all seem plausible.

The best way to answer a multiple-choice question is to read the question and try and answer it *before* looking at the possible answers. You may need to do some calculations – space is provided on the question paper for rough working. If the answer you thought of or calculate is among the possible answers – job done! Just have a look at the other possibilities to convince yourself that you were right.

If the answer you thought of isn't there, look at the possible answers and try to eliminate wrong answers until you are left with the correct one.

You don't lose any marks by having a guess (if you can't work out the answer) – but you won't score anything by leaving the answer blank. If you narrow down the number of possible answers, the chances of having a lucky guess at the right answer will increase.

To indicate the correct answer, put a cross in the box following the correct statement. If you change your mind, put a line through the box and fill in your new answer with a cross.

How Science Works

The idea behind 'How Science Works' is to give you insight into the ways in which scientists work: how an experiment is designed, how theories and models are put together, how data is analysed, how scientists respond to factors such as ethics and so on.

Many of the HSW criteria require practical or investigative skills and will be tested as part of your assessed practical work. However, there will be questions on the written units that cover all the HSW criteria. Some of these questions will involve data or graph interpretation, including the possible physical significance of the area between a curve and the horizontal axis, and determining quantities (with appropriate physical units) from the gradient and intercept of a graph (including at A2, log-linear and log-log graphs).

Another common type of HSW question will be on evaluating various steps in an experiment. For example,

- explain or justify why a particular piece of apparatus is used
- identify possible sources of systematic or random error
- explain why we use an instrument in a particular way
- what safety precautions would be relevant, and why?

You may be asked questions involving designing an investigation: these are likely to involve pieces of familiar practical work.

Other HSW questions may concentrate on issues surrounding the applications and implications of science (including ethical issues), or on using a scientific model to make predictions.

Extended questions

In the A2 units – Unit 4 and Unit 5 – you will come across questions with larger numbers of marks, perhaps up to 6 marks in one part of one question.

Questions in these Units are designed to be synoptic – in other words, they are designed for you to show knowledge gained in the earlier units. Bear this in mind when you answer the question: try to include relevant knowledge from your AS course when answering these questions.

Remember, too, that if part of a question is worth 6 marks, you need to make six creditworthy points. Think about the points that you will make and put them together in a logical sequence when you write your answer. On longer questions, the examiners will be looking at your QWC (Quality of Written Communication) as well as the answer you give.

Momentum and force

Momentum

All moving masses possess a quantity called **momentum**.

$$p = mv$$

where p is the momentum ($kg\,m\,s^{-1}$), m the mass (kg) and v the velocity ($m\,s^{-1}$). Although the base units for p are $kg\,m\,s^{-1}$, since $N \equiv kg\,m\,s^{-2}$ we often use Ns for the units of p.

Momentum is a **vector quantity** whose direction is the same as the velocity of the object. When we combine momentum values we need to draw a **vector triangle** to find the resultant.

ResultsPlus Watch out!

Remember that in a 'show that…is about…' question you should quote your answer to one more significant figure than the value given in the question.

Worked Example

A zebra can reach speeds up to $15\,m\,s^{-1}$. Show that the momentum of a zebra of mass 300 kg when running at top speed is about 5×10^3 Ns.

$p = mv = 300\,kg \times 15\,m\,s^{-1} = 4.5 \times 10^3\,Ns$

Worked Example

A raindrop of mass 0.065 g is falling through the air at its terminal velocity of $9\,m\,s^{-1}$. A gust of air moves it sideways, giving it a horizontal velocity component of $12\,m\,s^{-1}$. Calculate its momentum.

Calculate the resultant velocity of the raindrop.
Draw a sketch to help you.

$v^2 = 9^2 + 12^2$ $\tan\theta = \frac{12}{9}$

$v = \sqrt{81 + 144}$ $\theta = \tan^{-1} 1.33 = 53°$

$v = 15\,m\,s^{-1}$

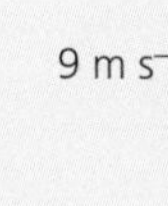

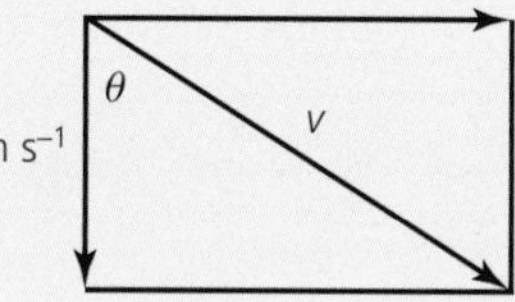

Convert mass to kg:

$m = 0.065\,g = 6.5 \times 10^{-5}\,kg$

Substitute values and solve:

$p = mv = 6.5 \times 10^{-5}\,kg \times 15\,m\,s^{-1} = 9.8 \times 10^{-4}\,Ns$

in a direction of 53° to the vertical.

ResultsPlus Watch out!

Many students don't draw and label a vector diagram; a clear diagram will ensure that you choose the correct values to use when working out the tan of the angle.

Newton's second law of motion

Newton formulated his second law of motion in terms of momentum changes.

A resultant force acting on an object causes a change of momentum. The rate of change of momentum is proportional to the magnitude of the resultant force and takes place in the direction of the resultant force.

Written as an equation:

$$F = \frac{\Delta(mv)}{\Delta t}$$

where F is the resultant force (in N) and Δt is the time (in s) for which the force acts.

- Δ (delta) means 'change in'.
- If mass is constant, $\Delta(mv) = m(v - u)$. Since $a = (v - u)/\Delta t$, the relationship is equivalent to $F = ma$.

Worked Example

A car of mass 1200 kg is travelling at 30 m s^{-1} just before crashing into a stationary barrier. If the car is brought to rest in 0.1 s, calculate the size of the average force that the car exerts on the barrier during the collision.

$F = \frac{\Delta(mv)}{\Delta t}$ gives the force *on the car.*

Substitute values and solve.

The car is brought to rest, so the final velocity is 0.

$$\therefore F = \frac{1200\,\text{kg} \times (0 - 30)\,\text{m s}^{-1}}{0.1\,\text{s}} = -3.6 \times 10^5\,\text{N}$$

The '–' indicates that the force on the car is backwards.

So the average force that the car exerts *on the barrier* is also 3.6×10^5 N, in the direction that the car was originally travelling.

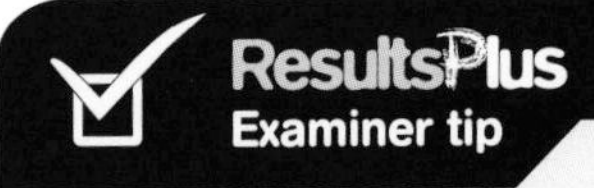

It is good practice to give your answer to a calculation to the same number of significant figures as the values given in the question.

Collision forces

Re-arranging the equation from Newton's second law gives:

$$F \times \Delta t = \Delta(mv)$$

The product of force and time for which the force acts is sometimes referred to as the **impulse**. From Newton's second law, impulse is equal to the change in momentum.

In order to achieve a change in momentum, a resultant force must act on the object. The longer time for which the force acts, the smaller the force needed for a given change in momentum. In car design one aim is to engineer a long impact time so that any collision force is reduced.

Quick Questions

Q1 A bungee jumper of mass 75 kg uses a bungee cord of length 25 m when jumping from a high bridge. Calculate the momentum of the jumper at the instant when the cord begins to stretch. What assumption have you made?

Q2 A car of mass 800 kg decelerates from 30 m s^{-1} to 10 m s^{-1} as it approaches a junction. What is its change in momentum?

Q3 An ice-hockey puck has a momentum of 1.2 N s just before making impact with the boards at the edge of the rink. It rebounds with a momentum of 1.0 N s after making contact for 0.15 s. Calculate the average force acting on the puck during the collision.

High-speed trains are designed with crumple zones to reduce the forces that would act during a collision. Use Newton's second law to explain how crumple zones achieve this.

Momentum and collisions

Conservation of momentum

In any collision in which no external forces act, the total momentum remains constant. This is referred to as the **principle of conservation of momentum**.

If two bodies, A and B, collide with no forces other than the collision forces (e.g. negligible friction), then:

$$(\text{momentum of A})_1 + (\text{momentum of B})_1 = (\text{momentum of A})_2 + (\text{momentum of B})_2$$

where 1 and 2 refer to the situation *just before* and *just after* the collision.

Since momentum is a vector quantity its direction must be taken into account. For bodies moving in the same straight line, it may be positive or negative depending upon their direction of motion.

Worked Example

Two skaters, one of mass 75 kg and the other of mass 55 kg, are travelling together across the ice in a straight line at a speed of $3.5\,\text{m s}^{-1}$. The skaters push away from each other along the line in which they are travelling, and the heavier skater moves backwards at $1.5\,\text{m s}^{-1}$. What is the other skater's new velocity?

$$(\text{momentum of A})_1 + (\text{momentum of B})_1 = (\text{momentum of A})_2 + (\text{momentum of B})_2$$

Rearrange and substitute:

$$v = \frac{(75\,\text{kg} \times 3.5\,\text{m s}^{-1}) + (55\,\text{kg} \times 3.5\,\text{m s}^{-1}) = 75\,\text{kg} \times (-1.5\,\text{m s}^{-1})}{55\,\text{kg}}$$

$$v = \frac{567}{55}\,\text{m s}^{-1} = 10.3\,\text{m s}^{-1}$$

So the lighter skater's velocity increases to $10\,\text{m s}^{-1}$ in the original direction.

Remember to indicate directions by using a sign convention. In this example 'forwards' is positive and 'backwards' negative.

Energy in collisions

In **elastic collisions** the total **kinetic energy** E_k is conserved (it is the same before and after the collision). If kinetic energy is not conserved the collision is referred to as an **inelastic collision**. The 'missing' energy has become some other form, such as thermal energy. When objects stick together a large fraction of the initial kinetic energy is usually converted to other forms.

Note that, since $p = mv$,

$$p^2 = m^2v^2 \therefore \frac{p^2}{2m} = \frac{mv^2}{2} = E_k$$

So

$$E_k = \frac{p^2}{2m}$$

Oblique (glancing) impacts

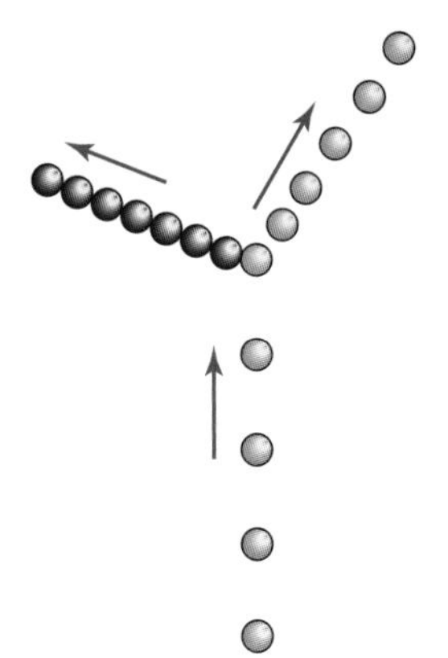

Collisions do not always take place along a straight line. In such collisions we resolve momentum into two perpendicular directions and apply the principle of conservation of momentum to each direction.

For elastic collisions between two *equal mass objects*, one of which is initially at rest, we can write:

$$\frac{p^2}{2m} = \frac{p_1^2}{2m} + \frac{p_2^2}{2m}$$

$$\therefore p^2 = p_1^2 + p_2^2$$

For momentum conservation ($p = p_1 + p_2$), the objects must move at right angles to each other after the collision, otherwise the two equations cannot be consistent.

Worked Example

In a game of snooker the cue ball makes a glancing impact with a blue snooker ball (both of mass 150 g). After the impact the two balls move at right angles to each other. The blue snooker ball moves at 3.00 m s^{-1} at an angle of 30° to the original direction of the cue ball. Calculate the velocity of the cue ball after the impact, assuming an elastic collision.

After the collision the cue ball moves at 60° to its initial direction.

Resolve final velocities perpendicular to initial direction of cue ball – see the diagram.

Before: u — After: 3 sin 30°, 3 m s^{-1}, 30°, 60°, v sin 60°, v

Consider initial and final momentum perpendicular to initial direction:

$$p_i = 0 \quad p_f = m \times 3\,\text{m s}^{-1} \times \sin 30° - mv\sin 60°$$

Apply the principle of conservation of momentum:

$$0 = m \times 3\,\text{m s}^{-1} \times \sin 30° - mv\sin 60°$$

$$v\sin 60° = 1.5\,\text{m s}^{-1}$$

$$\therefore v = 1.7\,\text{m s}^{-1}$$

ResultsPlus Examiner tip

Always draw a diagram to represent the situations before and after the collision. This will help to ensure that you write down the momentum expressions correctly.

Quick Questions

Q1 A car of mass 900 kg travelling at 25 m s^{-1} approaches a stationary van of mass 1600 kg. If the vehicles collide and move off together in the original direction of the car, what is their combined speed immediately after the collision?

Q2 A large ball of mass 1.05 kg travelling at 5 m s^{-1} hits a stationary smaller ball of mass 666 g. The large ball continues in the same direction after the impact at a speed of 1.1 m s^{-1}. The small ball continues in the same direction.
a What is the small ball's speed?
b Is the collision elastic or inelastic?

Q3 A bullet of mass 50 g is fired horizontally and embeds itself into a stationary block of wood of mass 1.25 kg. Calculate the speed of the bullet if the block moves forward with a speed of 3.9 m s^{-1} after the impact.

Thinking Task

Two cars collide at an intersection between two roads. One car has a mass of 900 kg and is travelling at due North at 25 m s^{-1}, and the other car has a mass of 750 kg and is travelling due East at 35 m s^{-1}. The cars remain together after the collision. Calculate the velocity of the cars immediately after the collision, and the fraction of kinetic energy lost. Account for the loss of kinetic energy.

Circular motion 1

Angular velocity

We can describe the speed of an object moving in a circle by its speed *around* the circle, as

$$v = \frac{2\pi r}{T}$$

where r is the radius of the circle and T is the time taken for one complete revolution.

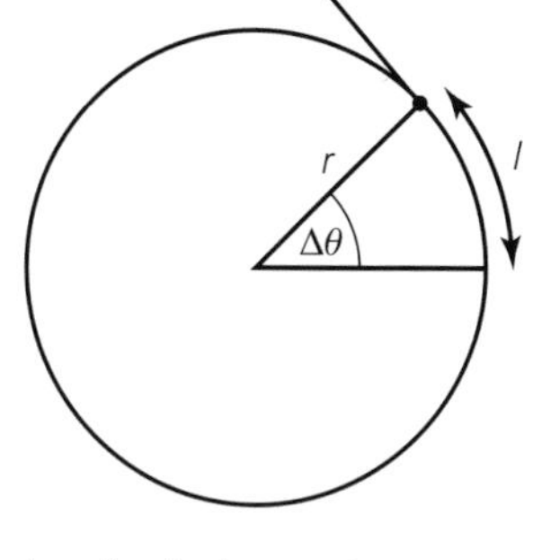

Angular displacement

We could also consider the **angular displacement**, $\Delta\theta$, that the object goes through as it moves around the circular path. This is measured relative to the centre of the circle.

$\Delta\theta$ is measured in **radians**. The angle θ in radians is defined as $\theta = l/r$ where l is the arc length that gives the angle θ at the centre of the circle of radius r. Since 2π radians make a complete circle (360°), to convert an angle in degrees into radians you divide the angle by 360° and multiply by 2π.

The **angular velocity**, ω, is given by:

$$\omega = \frac{\Delta\theta}{\Delta t}$$

where $\Delta\theta$ is the angular displacement in a time Δt. Angular velocity has units $\text{rad}\,\text{s}^{-1}$.

The tangential velocity and the time for one revolution can both be expressed in terms of angular velocity:

$$v = \omega r \quad \text{and } T = \frac{2\pi}{\omega}$$

Worked Example

a A fairground carousel takes 12 s to make one complete revolution. What is the angular velocity of the carousel?

b There are three rows of horses on the carousel – the inside row has a radius of 4 m and the outside row has a radius of 8 m. Compare the speeds of riders on the outside to those on the inside.

---

a $\omega = \frac{2\pi}{T} = \frac{2\pi}{12\,\text{s}} = 0.52\,\text{rad}\,\text{s}^{-1}$

b $v_o = \omega r_o \quad v_i = \omega r_i$

$$\frac{v_o}{v_i} = \frac{r_o}{r_i} = \frac{8}{4} = 2$$

So riders on the outside have a speed double that of those on the inside.

Centripetal acceleration

When an object moves in a circular path it is changing its direction and therefore it is accelerating. If the object is moving with a constant speed then the acceleration must be directed towards the centre of the circular path, perpendicular to its tangential motion. This is referred to as the **centripetal acceleration**, a.

$$a = \frac{v^2}{r}$$

where r is the radius and v is the tangential velocity around the circle.

Since $v = \omega r$, we can rewrite this expression for the centripetal acceleration as:

$$a = \frac{(\omega r)^2}{r} = \frac{\omega^2 r^2}{r} = \omega^2 r$$

Worked Example

A training centrifuge is able to spin trainee astronauts in a circular path of diameter 17.0 m. An astronaut takes 4.10 seconds to make one complete revolution. When riding the centrifuge, what is the astronaut's velocity around the circle, and what is his centripetal acceleration as a multiple of g?

$$v = \frac{2\pi r}{T} = \frac{2 \times \pi \times 8.5\,\text{m}}{4.1\,\text{s}} = 13.0\,\text{m s}^{-1}$$

$$a = \frac{v^2}{r} = \frac{(13.0\,\text{m s}^{-1})^2}{8.5\,\text{m}} = 19.9\,\text{m s}^{-2} = 2g$$

Quick Questions

Q1 A circular space station of radius 1000 m is designed to rotate about an axis through its centre to simulate gravity. How many revolutions per minute must it perform in order to produce an artificial gravity of 9.80 m s^{-2}?

Q2 Communications satellites orbit above the equator in orbits of radius 4.24×10^7 m with a period of 24 hours. Calculate the angular velocity of a communications satellite and the centripetal acceleration that it experiences.

Q3 A ceiling fan is turning at a rate of 100 revolutions per minute. A spider is clinging to a blade of the fan. If the spider experiences a centripetal acceleration greater than $0.3g$, it will lose its grip on the blade and be flung off. How far from the centre of the fan can the spider safely go?

ResultsPlus Watch out!

Remember to calculate the radius from the diameter before substituting.

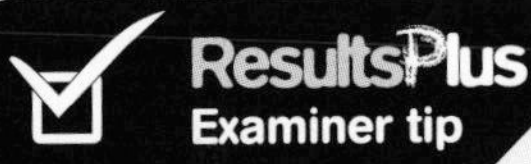

ResultsPlus Examiner tip

Watch out for units. Sometimes some quantities may be given in non-standard units – always make sure that you convert quantities into SI units *before* substituting into formulas.

Thinking Task

Telstar, the first communications satellite, was placed in an orbit 3300 km above the surface of the Earth and orbited the Earth once every 157 minutes. Calculate the satellite's

a angular velocity,
b acceleration.

The radius of the Earth is 6400 km.

Circular motion 2

Centripetal force

A body moving in a circular path is accelerating, so there must be a resultant force acting on it. This acts towards the centre of the circle. We call this the **centripetal force**, F. Applying Newton's second law ($F = ma$), we can write:

$$F = \frac{mv^2}{r}$$

The centripetal force must arise from external forces that act on the object. For example, if a ball on a string is whirled in a horizontal circle the centripetal force is provided by the tension in the string.

Worked Example

An amusement park 'loop the loop' ride has a loop of radius 8.0 m. Cars go around the inside of the loop, having descended from a vertical height of 7 m above the top of the loop.

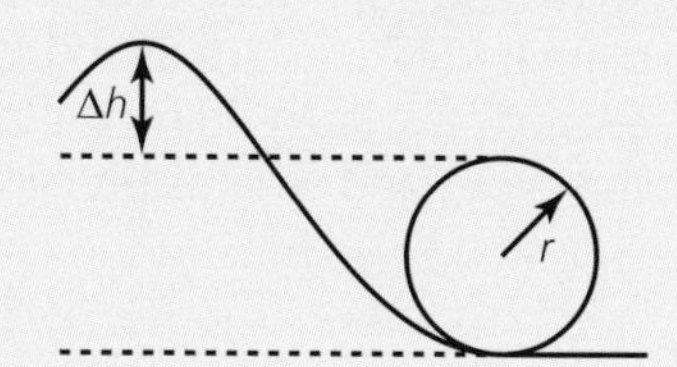

a Calculate the speed of the cars when they are at the top of the loop. Hence calculate the centripetal acceleration of the cars at the top of the loop.

b Joe (mass 65 kg) is on the ride. What is the force exerted on Joe by the car when the car is at the top of the loop?

---

a Use energy conservation to calculate the speed at the top of the loop.

$$\frac{1}{2}mv^2 = mg\Delta h$$

$$\therefore v^2 = 2g\Delta h$$

$$v = \sqrt{2 \times 9.8\,\mathrm{m\,s^{-2}} \times 7\,\mathrm{m}} = 11.7\,\mathrm{m\,s^{-1}}$$

Now use $a = v^2/r$ to calculate the centripetal acceleration at the top of the loop:

$$a = \frac{(11.7\,\mathrm{m\,s^{-1}})^2}{8.0\,\mathrm{m}} = 17.1\,\mathrm{m\,s^{-2}}$$

b Apply Newton's second law and call the force R.

$$mg + R = ma$$

$$\therefore R = m(a - g)$$

$$= 65\,\mathrm{kg} \times (17.1 - 9.8)\,\mathrm{m\,s^{-2}}$$

$$= 475\,\mathrm{N}$$

Worked Example

An aeroplane of mass 1.25×10^5 kg is turning horizontally in the air, following a circular path of radius 15 km at a speed of $245\,\mathrm{m\,s^{-1}}$. Its centripetal acceleration is $4.0\,\mathrm{m\,s^{-2}}$ and the centripetal force on it is 5.0×10^5 N. In order to negotiate this path, the plane banks. Explain why this is necessary, and calculate a value for the angle of banking necessary.

The force needed to maintain a circular path is too great to be provided from air resistance. When the plane banks, there is a component of the lift force from the wings acting horizontally. It is this horizontal force that allows the plane to move in its path.

Draw a free-body diagram to represent the forces acting on the plane, and resolve the lift force L into horizontal and vertical components.

Apply Newton's first law to the vertical direction:

$$L\cos\theta = mg$$

Apply Newton's second law to the horizontal direction:

$$L\sin\theta = m\left(\frac{v^2}{r}\right)$$

Now divide the equations from the two steps above:

$$\frac{L\sin\theta}{L\cos\theta} = \frac{\left(\frac{mv^2}{r}\right)}{mg}$$

$$\therefore \tan\theta = \frac{v^2}{rg}$$

$$\theta = \tan^{-1}\frac{(245\,\mathrm{m\,s^{-1}})^2}{15\,000\,\mathrm{m} \times 9.8\,\mathrm{m\,s^{-2}}} = 22°$$

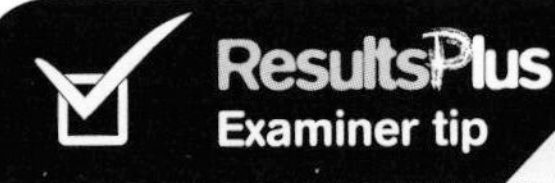

ResultsPlus Examiner tip

Don't try to remember formulas for banking – always work the problem through from first principles.

Quick Questions

Q1 A racing car of mass 500 kg goes around a circular racetrack with a diameter of 350 m at a speed of $150\,\mathrm{km\,h^{-1}}$. What is the centripetal force acting on the car?

Q2 A car of mass 1600 kg takes a bend of radius of curvature 30 m on a horizontal road at a constant speed.

- **a** Explain why there must be a resultant force on the car, even though its speed is constant.
- **b** In which direction does the frictional force act?
- **c** If the maximum frictional force from the road is 65% of the car's weight, what is the greatest speed at which the car can take the bend without skidding?

Q3 A skateboarder, of mass 65 kg, skates down and up a half-pipe (a semi-circular ramp) with a radius of 3 m. If he is travelling at $3.6\,\mathrm{m\,s^{-1}}$ when he is at the bottom of the half-pipe, what will be the total upward force at this point?

Thinking Task

Some trains, such as the Pendolino, tilt when going around curves. Consider such a train rounding a curve of radius 650 m at a speed of $120\,\mathrm{km\,h^{-1}}$.

- **a** Calculate the frictional force needed to keep a passenger of mass 75 kg in the seat if the train does not tilt.
- **b** Calculate the angle of tilt needed if there is to be no need of a frictional force from the seat.

Section 1: Further mechanics checklist

By the end of this section you should be able to:

Revision spread	Checkpoints	Spec. point	Revised	Practice exam questions
Momentum and force	Use the expression $p = mv$.	73	☐	☐
	Relate net force to rate of change of momentum in situations where the masses of the objects do not change (this is Newton's second law of motion).	75	☐	☐
Momentum and collisions	Apply the principle of conservation of linear momentum to problems involving objects moving along a straight line.	74	☐	☐
	Derive and use the expression $E_k = p^2/2m$ for the kinetic energy of particles (for particles moving much more slowly than the speed of light).	76	☐	☐
	Use data to calculate the momentum of particles.	77	☐	☐
	Apply the principle of conservation of linear momentum to problems in one and two dimensions (so you need to be able to resolve forces and speeds, using trigonometry).	77	☐	☐
	Explain the principle of conservation of energy.	78	☐	☐
	Use the principle of conservation of energy to determine whether a collision is elastic or inelastic.	78	☐	☐
Circular motion 1	Express angular displacement in radians and in degrees, and convert between those units.	79	☐	☐
	Explain the concept of angular velocity.	80	☐	☐
	Use the relationships $v = \omega r$ and $T = 2\pi/\omega$.	80	☐	☐
	Derive and use the expressions for centripetal acceleration $a = v^2/r$ and $a = r\omega^2$.	82	☐	☐
Circular motion 2	Explain that a resultant force (centripetal force) is required to produce and maintain circular motion.	81	☐	☐
	Use the expression for centripetal force $F = ma = mv^2/r$.	82	☐	☐

ResultsPlus
Build Better Answers

a State the principle of conservation of linear momentum. **[2]**

Examiner tip

Exam papers often ask you to state a law or principle – so it is worth memorising these!

Student answer	Examiner comments
In a system of interacting masses momentum stays constant as long as no external forces act.	This answer would score 1 mark, as it doesn't specify that the *total* momentum stays constant.

b A teacher demonstrates the conservation of momentum using a collision between a moving and a stationary trolley. Both trolleys stick together after the collision, and she measures the velocities using a motion sensor connected to a data logger. Explain why the velocities before and after the collision must be constant if the principle is to be convincingly demonstrated. **[2]**

Edexcel June 2008 Unit Test 1

Student answer	Examiner comments
This is because the momentum remains constant. Since the mass does not change, the velocity must not change ($p = mv$).	This answer would score no marks. ■ A **basic answer** would point out that conservation of momentum only applies if there are no external forces on the system (for 1 mark). ▲ An **excellent answer** would then say that if the speed was changing there would be external forces acting.

c Two students are watching an action film in which a car drives up a ramp onto the back of a moving lorry. Both vehicles are moving at high speed, the car slightly faster than the lorry. One student says this is impossible, because the car would not be able to stop before hitting the cab of the lorry.

The car has a mass 1250 kg and is moving at a speed of 28.0 m s^{-1}. The lorry has a mass 3500 kg and a speed of 25.5 m s^{-1}. The back of the lorry allows a braking distance of 5.0 m. By considering both momentum and energy show that the stunt is possible, provided a minimum force of about 600 N slows the car down. Treat the situation as one in which two objects join together, and support your explanations with calculations. **[8]**

Edexcel June 2008 Unit Test PSA 4

Examiner tip

To score full marks you will need to perform calculations as well as explaining your reasoning carefully.

Student answer	Examiner comments
Momentum is conserved so $v = \frac{(1250\,\text{kg} \times 28.0\,\text{m s}^{-1}) + (3500\,\text{kg} \times 25.5\,\text{m s}^{-1})}{(1250 + 3500)\,\text{kg}}$ $v = \frac{124\,250}{4750}\,\text{m s}^{-1} = 26.2\,\text{m s}^{-1}$ The car slows down from 28.0 m s^{-1} to 26.2 m s^{-1}, so $a = \frac{(26.2\,\text{m s}^{-1} - 28.0\,\text{m s}^{-1})}{5\,\text{s}} = -0.36\,\text{m s}^{-2}$ Apply : $F = ma$: $F = 1250\,\text{kg} \times 0.36\,\text{m s}^{-2} = 450\,\text{N}$ This is about 600 N.	This answer would score only 3 out of 8 marks. The student has calculated momentum values and worked out a correct value for the final velocity of the car. However the calculation of acceleration is *not* correct. The student has ignored the instruction in the question to consider energy as well as momentum. Another mark would have been gained just by giving the formula for kinetic energy. After calculating the kinetic energy lost in the collision, $W = Fs$ would need to be applied.

Practice exam questions

1 A runner of mass 65 kg is running around a circular running track at a speed of 5.0 m s^{-1}. The magnitude of the runner's change in momentum from starting to end position after half a revolution is

A 0 N s
B 325 N s
C 460 N s
D 625 N s **[1]**

2 When a stationary object explodes into a number of pieces

A both kinetic energy and momentum are conserved
B neither kinetic energy nor momentum is conserved
C only kinetic energy is conserved
D only momentum is conserved **[1]**

3 The drum of a washing machine, of diameter 35 cm, spins at 1500 rpm. The angular velocity of the drum is

A 8.8 rad s^{-1}
B 25 rad s^{-1}
C 55 rad s^{-1}
D 157 rad s^{-1} **[1]**

4 A fairground ride spins riders, in chairs hung from chains, into a horizontal circle. The chains make an angle to the vertical when the ride starts to rotate. This is

A because a centrifugal force acts outwards
B because a centripetal force acts outwards
C to allow a component of the tension in the chain to act inwards
D to allow a component of the riders' weight to act inwards **[1]**

5 A boy of mass 60 kg, standing on a skateboard, throws a load of mass 2.5 kg with a horizontal velocity of 12.5 m s^{-1}.

a Explain why the boy moves in the opposite direction to the load. **[2]**
b Calculate his velocity. **[2]**

6 A 'spud gun' uses compressed air to fire potato pellets at low velocity. A spud gun of mass 350 g fires a potato pellet of mass 2.5 g at a velocity of 12 m s^{-1}. What is the recoil velocity of the gun? **[4]**

7 Calculate the momentum of a bowling ball of mass 6.25 kg sent at 5.5 m s^{-1} towards a skittle of mass 1.2 kg. The bowling ball makes a head-on elastic collision with the skittle. What is the velocity of the skittle after the impact? **[6]**

8 An α-particle of mass 6.7×10^{-27} kg rebounds after making a head-on impact with a stationary nitrogen nucleus of mass 4.64×10^{-26} kg. The speed of the α-particle before the collision is 1.2×10^{7} m s^{-1} and after the collision it has a speed of 9.0×10^{6} m s^{-1}.

a Draw a diagram, showing the situation just before and just after the collision. **[1]**
b Calculate the speed of the nitrogen nucleus after the collision. **[3]**
c State whether the collision is elastic or inelastic, justifying your choice with a calculation. **[3]**

9 A car crashes into a wall at 25 m s^{-1}. The driver, who has a mass of 65 kg, is wearing a seatbelt, and is brought to rest in 75 ms.

a Calculate the momentum of the driver before the crash. **[2]**
b Calculate the average resultant force exerted on the driver during the impact. **[2]**
c Explain why the average resultant force is not the same as the force on the driver from the seatbelt. **[1]**

10 In a washing machine, clothes are placed inside a perforated metal drum. When the drum rotates at high speed, wet clothes are spun so that water escapes through the holes. The drum has a radius of 0.25 m and makes 900 revolutions per minute.

- **a** Show that the speed of the rim of the drum is approximately $20\,\text{m}\,\text{s}^{-1}$. **[2]**
- **b** Calculate the acceleration of a shirt in contact with the rim of the drum, and state its direction. **[2]**
- **c** What is the origin of the force that accelerates the shirt? **[1]**
- **d** Explain how water is 'spun out of' the clothes in the drum as it rotates. **[3]**

11 The 'chair-o-plane' is a fairground ride consisting of seats that are supported by chains hung from a circular structure. The circular structure rotates and the chairs swing out as they move in a circular path.

- **a** With the aid of a free-body diagram for a chair plus rider, explain why the chairs swing out as the structure rotates. **[2]**
- **b** What angle to the verticle will be formed by the chain of a chair-o-plane seat that is occupied by a 45 kg child, if the tension in the chain is 550 N? **[2]**
- **c** A nearby chair is occupied by a 90 kg man. Explain why the angle formed by the chain will be the same as for the child. **[2]**
- **d** What is the speed of the chair and rider, if they are moving in a circle of radius 10 m? **[2]**
- **e** Explain what happens to the angle made by the chain to the vertical as the ride speeds up. **[3]**

12 Two ice skaters, each of mass 65.0 kg, grab hands and spin in a circle once every 4.5 s. Their arms are 0.72 m long.

- **a** What is their angular velocity? **[1]**
- **b** How big is their centripetal acceleration? **[1]**
- **c** How hard are they pulling on one another? **[2]**
- **d** The skaters let go of each other whilst in the spin. Describe their subsequent motion in as much detail as you can. **[2]**

Electrostatics

Coulomb's law

The electrostatic force between two charged spherical objects obeys an **inverse square law**.

$$F = \frac{kQ_1Q_2}{r^2}$$

where Q_1 and Q_2 are the two charges, r is the separation of the centres of the two charged objects and k is a constant dependent upon the medium between the charges. For charges in a vacuum,

$$k = \frac{1}{4\pi\varepsilon_0}$$

where $\varepsilon_0 = 8.85 \times 10^{-12}\,\text{F m}^{-1}$, so

$k = 8.9 \times 10^{9}\,\text{m F}^{-1}$

An equal magnitude force acts on each charged object. The forces are repulsive for two like charges and attractive for two opposite charges. The forces form Newton's third law pair.

Remember to square the separation, r, when entering data into your calculator.

Worked Example

Geiger and Marsden investigated the deflection of alpha particles as they passed through thin gold foils. Calculate the force between a gold nucleus (charge $+79e$) and an alpha particle ($+2e$) when the separation between the gold nucleus and the alpha particle is $5.0 \times 10^{-14}\,\text{m}$. (The magnitude of the charge on an electron, e, is $1.6 \times 10^{-19}\,\text{C}$).

$$F = \frac{8.9 \times 10^{9}\,\text{m F}^{-1} \times (79 \times 1.6 \times 10^{-19}\,\text{C}) \times (2 \times 1.6 \times 10^{-19}\,\text{C})}{(5 \times 10^{-14}\,\text{m})^2} = 14\,\text{N}$$

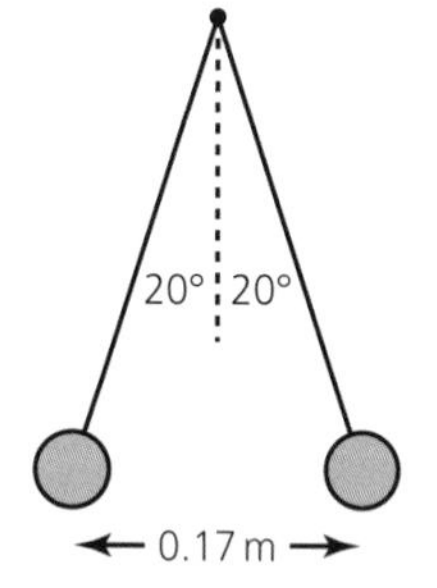

Worked Example

Two identical conducting spheres of mass $3 \times 10^{-6}\,\text{kg}$ are suspended from a fixed point by two threads. The spheres are initially in contact. Some charge is transferred onto one of the spheres, and the two spheres are seen to spring apart. At equilibrium the threads are at an angle of 20° to the vertical and the centres of the spheres are separated by a distance of 0.17 m.

a Explain why the spheres move away from one another.

b Calculate the charge on each sphere.

a The spheres share the charge equally. Like charges repel.

b Use $W = mg$ to calculate the weight of each sphere.

$$W = mg = 3 \times 10^{-6}\,\text{kg} \times 9.8\,\text{N kg}^{-1} = 2.9 \times 10^{-5}\,\text{N}$$

Use trigonometry to calculate the size of the repulsive force on each sphere.

$$\tan 20° = \frac{F}{W} \qquad \therefore F = W \tan 20° = 2.9 \times 10^{-5}\,\text{N} \times \tan 20° = 1.1 \times 10^{-5}\,\text{N}$$

Use Coulomb's law to calculate the charge on each sphere.

ResultsPlus Examiner tip

A2 questions are much less structured than those at AS. Think through what you might need to know and use the data given to calculate *something*, even if you are not sure how you will proceed.

$$F = \frac{1}{4\pi\varepsilon_0}\frac{Q_1Q_2}{r^2}$$

$$\therefore Q^2 = 4\pi\varepsilon_0 \times Fr^2$$

$$= 4\pi \times 8.85 \times 10^{-12}\,\text{F m}^{-1} \times 1.1 \times 10^{-5}\,\text{N} \times (0.17\,\text{m})^2$$

$$Q = \sqrt{1.1 \times 10^{-5}\,\text{N} \times 4\pi \times 8.85 \times 10^{-12}\,\text{F m}^{-1} \times (0.17\,\text{m})^2}$$

$$= 6.0 \times 10^{-9}\,\text{C}$$

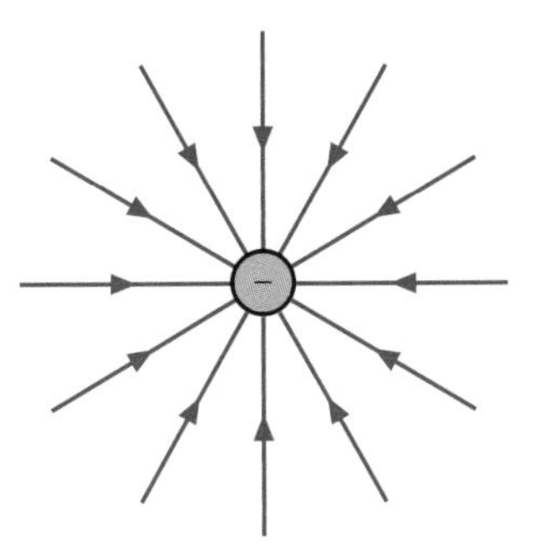

Electric field around a negative point charge

Radial electric fields

There is an **electric field** around every charged object. The interaction between electric fields produces electrostatic forces.

The **electric field strength**, E, at a point is the force, F, acting per unit charge, q, at that point:

$$E = \frac{F}{q}$$

E has units N C^{-1}. It is a vector quantity; its direction is the same as that of the force. The closer together the field lines, the stronger the field. Field lines point *towards* a negative charge.

The field around a negative point charge is therefore directed radially inwards towards the charge; around a spherical *positive* charge distribution it is directed radially outward.

The metal sphere acts as if all the charge were concentrated at its centre

We can calculate the field strength at a position in a radial field by using Coulomb's law. If the charge producing the field is Q and a small charge in the field is q:

$$E = \frac{F}{q} \text{ and } F = \frac{kQq}{r^2} \text{ so } E = \frac{kQ}{r^2}$$

Worked Example

A van de Graaff generator has a spherical metallic dome of diameter 50 cm. When the generator has been running for some time the electric field strength at the dome's surface is $400\,\text{N C}^{-1}$. Calculate the charge stored on the dome.

Rearrange and substitute into $E = \frac{kQ}{r^2}$:

$$Q = \frac{Er^2}{k}$$

$$= \frac{400\,\text{N C}^{-1} \times (0.25\,\text{m})^2}{8.9 \times 10^9\,\text{m F}^{-1}}$$

$$= 2.81 \times 10^{-9}\,\text{C}$$

ResultsPlus
Watch out!

Remember to halve the diameter to obtain the radius before you substitute.

Quick Questions

Q1 In the Bohr model of the atom, hydrogen consists of an electron orbiting a single proton. If the average separation of the proton and the electron is $5.28 \times 10^{-11}\,\text{m}$, calculate the size of the electrostatic force on the electron.

Q2 Calculate the force exerted on an alpha particle (charge $2e$) when it is placed in a field of strength $2 \times 10^6\,\text{N C}^{-1}$.

Q3 The electric field strength near to the surface of the Earth is $140\,\text{N C}^{-1}$. Calculate the net charge on the Earth, assuming that this charge is distributed symmetrically over the surface of a sphere of radius 6370 km.

Thinking Task

A small sphere carrying a charge of 400 pC is moved to various positions in a radial field set up by a point charge. The table shows some results that were obtained.

Distance from point charge / cm	Force on sphere / μN
1.00	90.0
2.00	
	30.0

a Copy and complete the table.

b Calculate the charge producing the radial field.

Capacitors

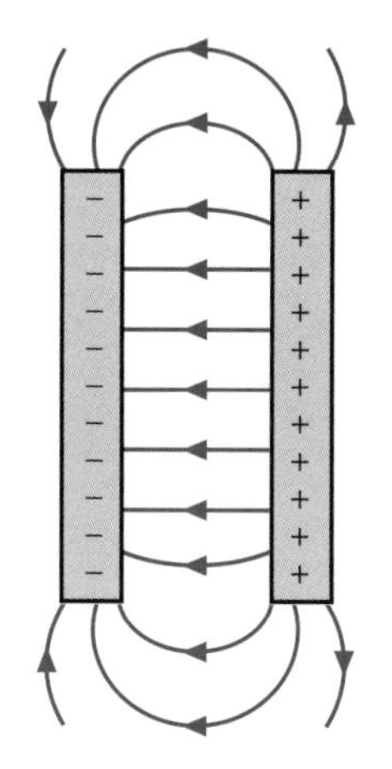

Electric field produced by two parallel plates

Uniform electric fields

Not all fields are radial. An example of a **uniform electric field** is found in the space between two parallel plates when there is a potential difference applied between the plates.

There is a uniform field between the plates, where the field lines are *parallel* and *equally spaced*.

The strength of the uniform field between the plates is given by:

$$E = \frac{V}{d},$$

where V is the potential difference between the plates and d is their separation. The units are V m^{-1}, which are equivalent to N C^{-1}.

Worked Example

The base of a thundercloud is 2.0 km above the surface of the Earth. The potential difference between the base of the cloud and the Earth's surface is 60 MV. What is the force on a free electron in the space between the cloud and the ground?

Use $E = V/d$ to calculate the electric field strength:

$$E = \frac{6 \times 10^7\,\text{V}}{2 \times 10^3\,\text{m}} = 3 \times 10^4\,\text{V m}^{-1}$$

Use $F = E \times Q$ to calculate the force on an electron:

$$F = 3 \times 10^4\,\text{V m}^{-1} \times 1.6 \times 10^{-19}\,\text{C} = 4.8 \times 10^{-15}\,\text{N}$$

Capacitors

A capacitor is a device that stores energy by separating charge. Placing charge on the capacitor results in a potential difference. The **capacitance**, C, of the system is defined as the charge stored, Q, per unit potential difference, V.

$$C = \frac{Q}{V}$$

Capacitance is measured in farads, F. $1\,\text{F} = 1\,\text{C V}^{-1}$. This is a huge unit; typical capacitances are in the range μF down to pF.

ResultsPlus
Watch out!

Capacitance has a symbol C but is measured in farads, F. Charge has a symbol Q but is measured in coulombs, C. Don't confuse the symbols and units.

Worked Example

A capacitor stores a charge of 0.063 μC when it is connected to a source of emf of 9 V. What is its capacitance? How much charge will it store for an applied potential difference of 15 V?

Use $C = Q/V$ to calculate the capacitance:

$$C = \frac{0.063 \times 10^{-6}\,\text{C}}{9\,\text{V}} = 7.0 \times 10^{-9}\,\text{F} = 7.0\,\text{nF}$$

Use $Q = C \times V$ to calculate the new charge stored:

$$Q = 7.0 \times 10^{-9}\,\text{F} \times 15\,\text{V} = 1.1 \times 10^{-7}\,\text{C} = 0.11\,\mu\text{C}$$

Charging and discharging

A potential difference builds up across a capacitor as it charges. The capacitor is fully charged once the potential difference across it becomes equal to the emf of the source. If we disconnect the capacitor from the source, we can use it to make a current flow around a circuit.

Energy is stored in the capacitor because work is done as charge moves through the net potential difference in the circuit. This becomes electrostatic potential energy (i.e. energy stored in the electric field between the capacitor plates).

The work done, W, is equal to the shaded area under the graph of potential difference between the plates, V, against charge on the plates, Q.

$$W = QV_{av}$$

where V_{av} is the *average* potential difference that the charge moves through. Since there is a proportional relationship between Q and V, the average potential difference is $V/2$, where V is the maximum potential difference. Hence the energy stored is given by:

$$W = Q \times \frac{1}{2}V$$

Since $Q = CV$, we can write two expressions for the energy stored:

$$W = \frac{1}{2}QV \text{ and } W = \frac{1}{2}CV^2$$

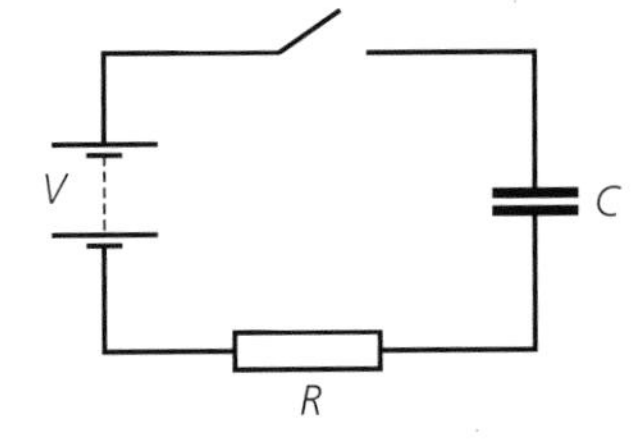

Simple circuit for charging a capacitor

Worked Example

The dome of a van de Graaff generator has a capacitance of 20 pF. Calculate the energy stored when the dome is raised to a potential of 100 kV.

$$W = \frac{1}{2}CV^2 = 0.5 \times 20 \times 10^{-12}\,\text{F} \times (100 \times 10^3\,\text{V})^2 = 0.1\,\text{J}$$

Quick Questions

Q1 In an experiment to determine the electronic charge, charged oil droplets are suspended in the electric field between two parallel plates. A droplet of mass 8.1×10^{-13} kg is held stationary between the parallel plates, of separation 1.6 cm, when a potential difference of 5100 V is set up between the plates. Calculate the charge on the droplet.

Q2 In order to ignite the air–fuel mixture in a combustion engine there needs to be an electric field of about $10\,\text{MV m}^{-1}$ in the gap in the spark plug. If the spark plug gap is 1.4 mm, calculate the potential difference required to produce the required field strength between the gap.

A spark plug

Q3 A van de Graaff generator can generate a voltage of 500 kV. Calculate the capacitance of the dome if it stores 7.5 J of energy when fully charged.

ResultsPlus Watch out!

In the graph of Q against V for a capacitor under charge, the potential difference the charge moves through decreases as the charge on the plates of the capacitor increases.

Thinking Task

EDL capacitors can bridge the gap between batteries and capacitors by providing energy bursts that neither batteries nor traditional capacitors can provide efficiently. One such capacitor with $C = 3.5$ F operates with a 4.6 V supply. Calculate the energy stored when this capacitor is fully charged. Explain why a capacitor is suitable for energy bursts, but unsuitable for sustained electrical power uses.

Exponential changes

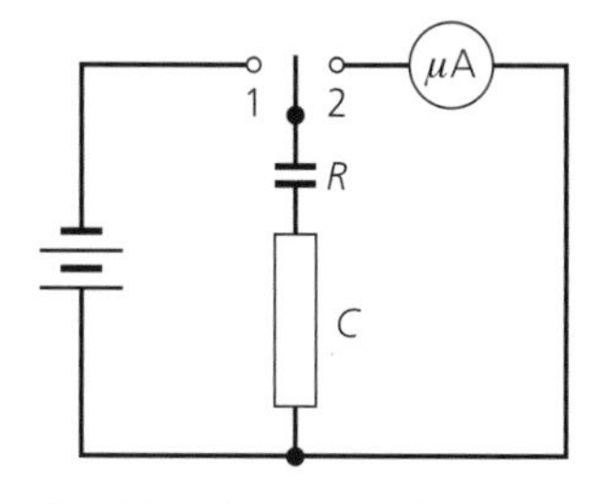

Circuit for charging and discharging a capacitor

Discharge of a capacitor

Moving the switch from 1 to 2 in the circuit shown begins the discharge of the capacitor. As the charge decreases, so does the potential difference across the capacitor and hence also the current.

The discharging process is an example of **exponential decay**, or a constant ratio change. The time taken for the charge on the capacitor to fall to a given fraction of its starting value is always the same for a given circuit. This depends upon the capacitance, C, and resistance, R, in the circuit.

$$Q = Q_0 e^{\frac{-t}{RC}}$$

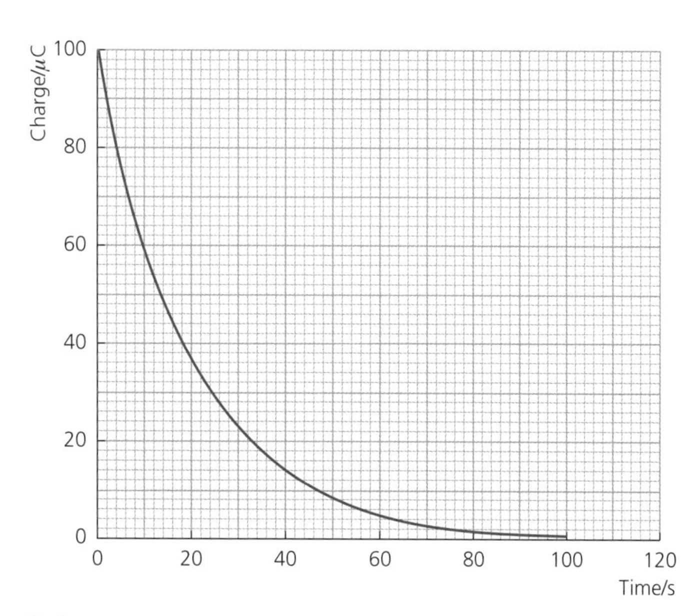

Discharge curve

Q_0 is the initial charge on the capacitor, Q is the charge remaining on the capacitor after a time t.

If we consider a time period $t = RC$, then $Q = Q_0/e = 0.37Q_0$. The value RC is referred to as the **time constant** of the circuit. In every period equal to the time constant, the charge falls to 37% of its value at the start of that period.

The charge remaining is very close to zero after a time of 5 × time constant, hence we often take $5RC$ as the time for the capacitor to discharge fully.

Since $V = Q/C$, the capacitor potential difference and so the discharging current also decrease exponentially.

When a capacitor is *charged* through a resistor, in one time constant the capacitor charge and pd reach 63% of their final values . Again, $5RC$ is often taken to be the time for the process to be complete.

To calculate the answer to one part of a question you sometimes need to refer to data provided or calculated in a previous part of the question.

Worked Example

A 68 μF capacitor is charged by a potential difference of 2000 V and then discharged through a circuit with resistance 60 Ω. How long will it take to discharge completely? What current will flow?

Find the charge when fully charged:

$$Q = CV = 68 \times 10^{-6}\,\text{F} \times 2000\,\text{V} = 0.14\,\text{C}$$

Find the time constant:

$$RC = 60\,\Omega \times 68 \times 10^{-6}\,\text{F} = 4.1 \times 10^{-3}\,\text{s}$$

Time to discharge completely = $5RC = 5 \times 4.1 \times 10^{-3}\,\text{s} = 2.0 \times 10^{-2}\,\text{s}$ or 20 ms

The average current will be:

$$I = \frac{Q}{t} = \frac{0.14\,\text{C}}{2.0 \times 10^{-2}\,\text{s}} = 7\,\text{A}$$

Worked Example

A student sets up the charging/discharging circuit shown opposite, to investigate the discharge of a capacitor of value 220μF. The capacitor is charged by moving the switch to position 1 and then discharged through the large-value resistor by moving the switch to position 2. The student reads the discharge current every 10 s. The results table is shown.

t/s	I/μA
0	100
10	68
20	47
30	32
40	22
50	15
60	10

Use a straight-line graph to determine the time constant of the circuit, and hence calculate a value for the resistance in the circuit.

The current–time relationship will be:

$$I = I_0 e^{-t/RC}$$

$$\therefore \log_e I = \log_e I_0 - \frac{t}{RC}$$

Hence a graph of $\log_e I$ against t should be a straight line with gradient $-1/RC$. Taking the natural log of the I values, the graph shown is obtained. The gradient is

$$\frac{(2.3 - 4.6)}{(60 - 0)\,\text{s}} = -0.0383\,\text{s}^{-1}$$

Since the gradient = $-1/RC$,

$$\therefore RC = -\frac{1}{\text{gradient}}$$

$$\therefore R = -\frac{1}{220 \times 10^{-6}\,\text{F} \times (-0.0383\,\text{s}^{-1})} = 1.2 \times 10^5\,\Omega$$

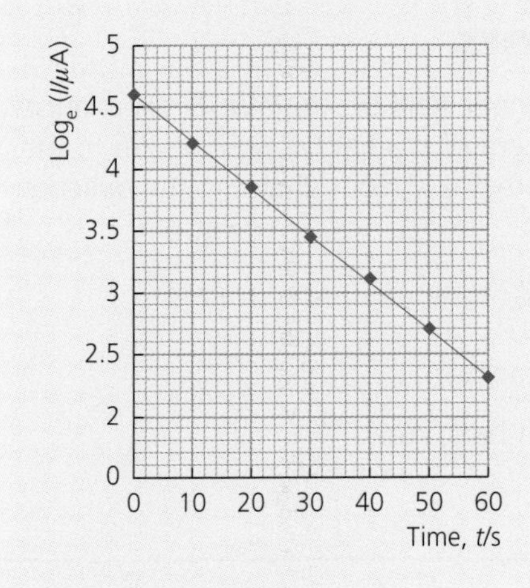

Quick Questions

Q1 A capacitor charges through a 0.22 MΩ resistor. Calculate the capacitance if the time taken for it to reach full charge is 5 minutes.

Q2 Show that the time constant RC has units of seconds.

Q3 A 47μF capacitor is charged to a potential difference of 15 V and then discharged through a 100 kΩ resistor. How long does it take for half of the energy stored in the capacitor to have dissipated in the circuit?

Thinking Task

Mains-operated power supplies have large-value capacitors to help keep the output voltage constant. In a full-wave rectified supply, without a capacitor the output falls to zero every 10 ms. If the output must be maintained at a value at least 90% of its maximum value when a load of 1.0 kΩ is connected to the output terminals, show that the minimum capacitance needed in the power supply circuit is about 100μF.

Magnetic fields

Magnetic flux

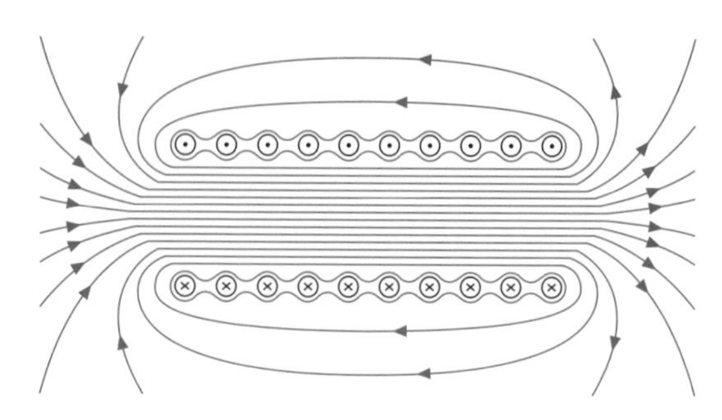

The magnetic field of a solenoid

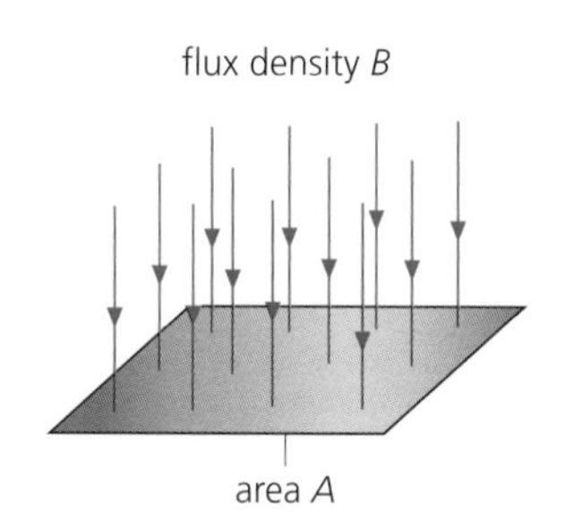

Flux $\Phi = B \times A$

A wire carrying a current has a circular magnetic field around it. When the wire is wound into a long solenoid there is a **uniform** field inside it – note the parallel, equally spaced field lines. Outside the solenoid the field is similar to that of a bar magnet.

Lines representing the magnetic field in a given region are called 'lines of **magnetic flux**'.

- By convention, arrows indicate the forces that would act on a magnetic north pole.
- The lines are continuous, and must not cross each other.
- The closer the lines, the stronger the magnetic field.

The number of lines passing through a unit area perpendicular to the field represents the **flux density**, B, and is a measure of the **magnetic field strength**.

Flux density, B, is measured in tesla, T. Flux, Φ, is measured in webers, Wb.

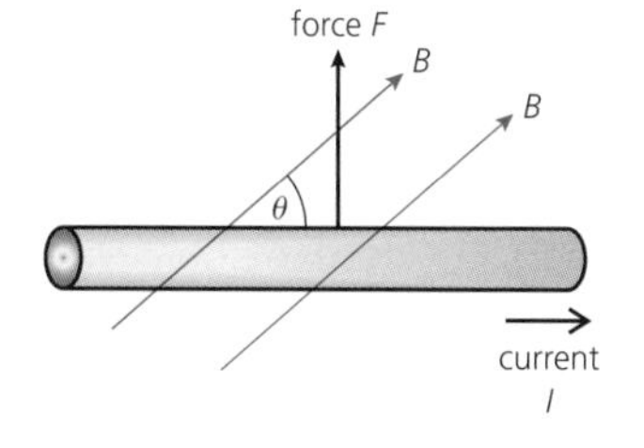

Current-carrying conductors

A length of wire, l, carrying a current, I, in a magnetic field of strength, B, experiences a magnetic force, F, as long as there is a component of the field at right angles to the current. The maximum force will be when the current is at right angles to the direction of the field.

$$F = BIl\sin\theta$$

where the angle θ is measured as shown in the diagram.

The direction of the force is given by Fleming's left hand rule.

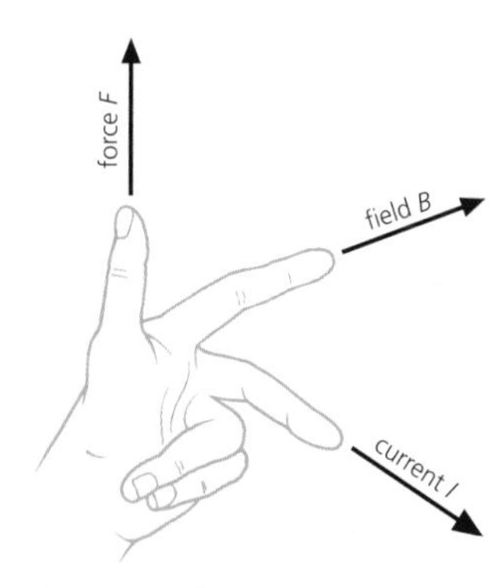

Fleming's left hand rule

Worked Example

A 4.0 cm length of wire passes between magnet poles. A current of 5.2 A flows perpendicular to the magnetic field. What is the field strength, if the force on the wire is 7.55×10^{-3} N?

Rearrange and substitute into $F = BIl\sin\theta$:

$$B = \frac{F}{Il\sin\theta}$$

$$= \frac{7.55 \times 10^{-3}\,\text{N}}{5.2\,\text{A} \times 4.0 \times 10^{-2}\,\text{m} \times 1} = 3.6 \times 10^{-2}\,\text{T}$$

Charged particle beams

A charge, Q, moving with speed, v, through a magnetic field, also experiences a magnetic force, F, as long as there is a component of the field perpendicular to the direction of charge movement.

The maximum force will be when the charge movement is at right angles to the direction of the field.

$$F = BQv\sin\theta$$

where the angle θ is measured as shown in the diagram.

ResultsPlus
Watch out!

In Fleming's left hand rule, **conventional current** is used, i.e. the direction of flow of *positive* charge. So if a question talks about electron movement, you must take I to be in the opposite direction to the electron movement.

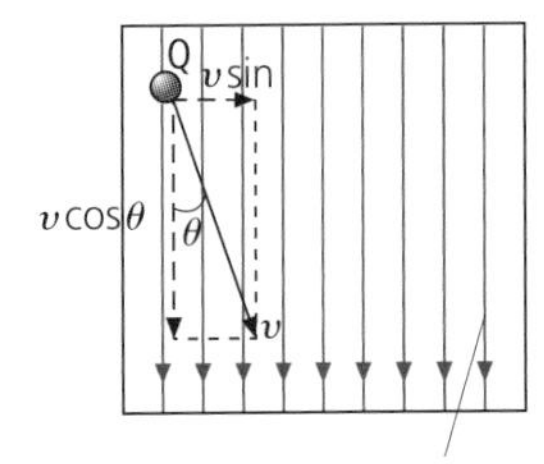

Worked Example

A fine beam tube contains a very low pressure gas so that the electron path can be seen. In such a tube, electrons are accelerated through a potential difference of 250 V in the electron gun.

a Show that electrons emerge from the gun with a speed of about $1 \times 10^7\,\mathrm{m\,s^{-1}}$, and hence calculate the magnetic force on them when they enter a magnetic field of strength $550\,\mu\mathrm{T}$, moving at right angles to the field.

b Explain why the electrons follow a circular path in the field, and calculate the radius of the path.

---

a Apply energy conservation to calculate the speed of the electrons.

gain in kinetic energy = work done by electric field in electron gun

$$\tfrac{1}{2}mv^2 = QV \quad \therefore v^2 = \frac{2QV}{m}$$

$$\therefore v = \sqrt{\frac{2 \times 1.6 \times 10^{-19}\,\mathrm{C} \times 250\,\mathrm{V}}{9.1 \times 10^{-31}\,\mathrm{kg}}} = 9.4 \times 10^6\,\mathrm{m\,s^{-1}}$$

Substitute values into $F = BQv$ to find the force:

$$F = 550 \times 10^{-6}\,\mathrm{T} \times 1.6 \times 10^{-19}\,\mathrm{C} \times 0.94 \times 10^7\,\mathrm{m\,s^{-1}}$$
$$= 8.3 \times 10^{-16}\,\mathrm{N}$$

b The magnetic force on the electrons acts at right angles to the plane containing B and v. Hence the force is always at right angles to the electron velocity, and produces a centripetal acceleration. Substitute values into the equation for centripetal force (see page 14), $F = mv^2/r$, and solve for r.

$$F = \frac{mv^2}{r} \quad \therefore r = \frac{mv^2}{F} = \frac{9.1 \times 10^{-31}\,\mathrm{kg} \times (0.94 \times 10^7\,\mathrm{m\,s^{-1}})^2}{8.3 \times 10^{-16}\,\mathrm{N}} = 0.097\,\mathrm{m}$$

ResultsPlus
Examiner tip

Quantities such as the charge and mass of an electron may not be given in the question – check the data sheet at the end of the exam paper!

Quick Questions

Q1 When sketching the magnetic field around a solenoid, a student draws two lines of the field that cross. Explain why this must be incorrect.

Q2 A section of power cable strung between two pylons 120 m apart carries a current of 1800 A. Calculate the force on the cable due to the Earth's magnetic field, which has a field strength component of $56\,\mu\mathrm{T}$ at right angles to the cable.

Q3 A beam of electrons enters a region of uniform magnetic field, moving at right angles to the field, and is deflected into a circular path. Draw a diagram to show the path of the electrons, including directions of the electron velocity v and the magnetic field B.

Thinking Task

An electron travelling at $2.7 \times 10^8\,\mathrm{m\,s^{-1}}$ is deflected into a circular path of diameter 3 cm when it enters a magnetic field of strength 0.24 T, at right angles to the field.

a Use this data to calculate the charge per unit mass, e/m, for the electron.

b The value for the charge-to-mass ratio at right angles to e/m for an electron is usually quoted as $1.76 \times 10^{11}\,\mathrm{C\,kg^{-1}}$. Comment on your answer in comparison to this textbook value.

Generating electricity

Electromagnetic induction

An emf is generated in any conductor that experiences a changing magnetic flux. The changing magnetic flux may be due to relative movement between the conductor and a magnetic field (the dynamo effect), or because the conductor is in a region of varying magnetic field strength (the transformer effect).

Flux $\Phi = B \times A\cos\theta$

For a conducting coil, the size of the effect increases with increasing turns of wire, so we refer to a change in **flux linkage**, $N\Phi$, where N is the number of turns. The flux through the coil depends upon the angle of the normal to the plane of the coil to the field direction, as shown in the diagram.

Faraday's law states that the magnitude of the emf, $\mathscr{E}$, is directly proportional to the rate of change of flux linkage.

$$\mathscr{E} \propto \frac{d(N\Phi)}{dt}$$

Lenz's law states that the induced emf must cause a current to flow in such a direction as to oppose the change in flux linkage that produces it, otherwise energy would appear from nowhere. Put together,

$$\mathscr{E} = -\frac{d(N\Phi)}{dt}$$

The direction of the induced emf can be found by the right hand rule.

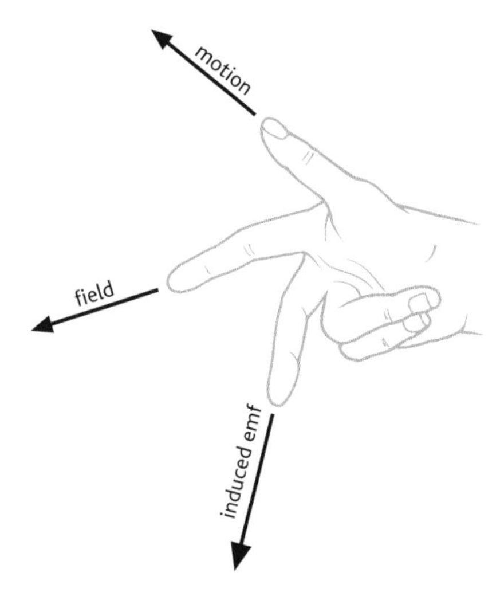

The right hand rule

Worked Example

A vertical car aerial is of length 1.2 m when fully extended. Show that the emf generated between its ends is about 1 mV when the car is moving at 35 m s^{-1} due east. Take the horizontal component of the Earth's magnetic field to be 22 μT.

Calculate the area swept out per second by the aerial:

$$\frac{\Delta A}{\Delta t} = \text{length of aerial} \times \text{speed of car}$$
$$= 1.2\,\text{m} \times 35\,\text{m s}^{-1} = 42\,\text{m}^2\,\text{s}^{-1}$$

Calculate the emf from the rate of change of flux linkage. Since $N\Phi = NAB$,

$$\frac{\Delta(N\Phi)}{\Delta t} = B \times \frac{\Delta A}{\Delta t}$$

$$\therefore \mathscr{E} = -\frac{\Delta(N\Phi)}{\Delta t} = -22 \times 10^{-6}\,\text{T} \times 42\,\text{m}^2\,\text{s}^{-1} = 9.2 \times 10^{-4}\,\text{V}$$

ResultsPlus Examiner tip

You may find it helpful to draw a sketch to help visualise a physical situation such as this.

The dynamo effect

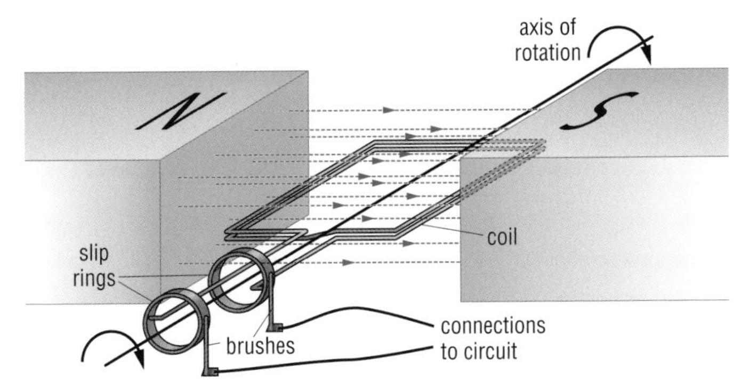

An AC generator

As the coil rotates, the magnetic flux linkage goes from zero (when the coil is parallel with the field) to a maximum (when the coil is perpendicular to the field). An alternating emf is generated.

The transformer effect

If the current in a coil changes, then so does the magnetic field produced. If the changing magnetic flux links with another coil then there will be an emf induced in that coil. The coil in which current is changing is referred to as the primary. The other coil is called the secondary.

To ensure maximum flux linkage between the two coils they are mounted on a soft iron core. This becomes magnetised when current flows in the primary, and flux is channelled to the secondary.

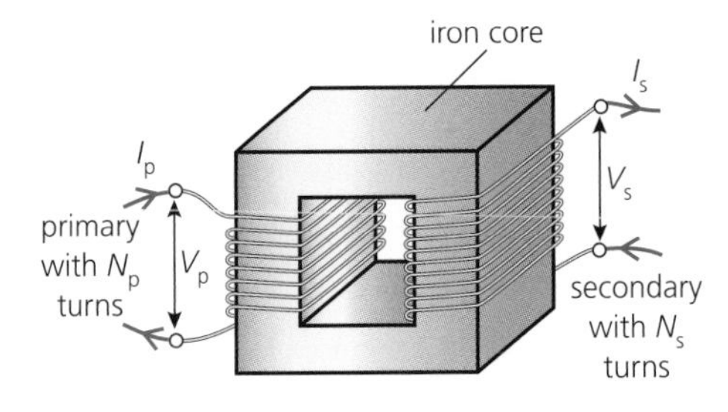

A transformer circuit. If $N_s > N_p$ then $V_s > V_p$ (and $I_s < I_p$)

Eddy currents

Any metal object placed in a region of changing magnetic field will have an emf induced in it. This emf causes current to circulate in the object. We call such currents eddy currents. The eddy currents in a transformer core dissipate energy ($P = I^2R$), so can lead to inefficiency. In situations where relative movement is involved, the eddy currents cause braking.

Quick Questions

Q1 A straight conductor of length 25 cm moves perpendicular to a magnetic field of strength 400 μT. If it moves 0.5 m in 2.5 s, what is the magnitude of the emf induced between its ends?

Q2 A metal propeller rotates in a vertical plane with an angular velocity of 85 rad s^{-1}. The length of the propeller from tip to tip is 2.5 m. Calculate the emf generated between the centre and the tip of the propeller. The horizontal component of the Earth's magnetic field strength is 22 μT.

Q3 A bar magnet dropped down a copper tube takes much longer to fall than a similarly sized unmagnetised steel bar of the same mass. Explain this observation.

Thinking Task

An electric toothbrush has a coil in the charger and one in the base of the brush. Explain how the battery in the toothbrush is able to be re-charged without any electrical contact between charger and toothbrush.

Section 2: Electric and magnetic fields checklist

By the end of this section you should be able to:

Revision spread	Checkpoints	Spec. point	Revised	Practice exam questions
Electrostatics	Explain what is meant by an electric field.	84		
	Recognise and use the expression for electric field strength $E = F/Q$.	84		
	Draw and interpret diagrams using lines of force to describe radial electric fields qualitatively.	83		
	Use the expression $F = kQ_1Q_2/r^2$, where $k = 1/4\pi\varepsilon_0$.	85		
	Derive and use the expression $E = kQ/r^2$ for the electric field due to a point charge.	85		
Capacitors	Recall that applying a potential difference to two parallel plates produces a uniform electric field in the central region between them.	86		
	Draw and interpret diagrams using lines of force to describe uniform electric fields qualitatively.	83		
	Recognise and use the expression $E = V/d$.	86		
	Use the expression $C = Q/V$.	87		
	Recognise and use the expression $W = \frac{1}{2}QV$ for the energy stored by a capacitor, and derive the expression from the area under a graph of potential difference against charge stored.	88		
	Derive and use related expressions, for example, $W = \frac{1}{2}CV^2$.	88		
Exponential changes	Recall that the growth and decay curves for resistor–capacitor circuits are exponential.	89		
	Know the significance of the time constant RC.	89		
	Use the expression $Q = Q_0e^{-t/RC}$, and derive and use related expressions for exponential discharge in RC circuits, for example, $I = I_0e^{-t/RC}$.	90		
Magnetic fields	Use the terms magnetic flux density B, flux Φ and flux linkage $N\Phi$.	91		
	Recognise and use the expression $F = BIl\sin\theta$.	92		
	Apply Fleming's left hand rule to currents.	92		
	Recognise and use the expression $F = Bqv\sin\theta$.	93		
	Apply Fleming's left hand rule to charges.	93		
Generating electricity	Explain qualitatively the factors affecting the emf induced in a coil when there is relative motion between the coil and a permanent magnet, and when there is a change of current in a primary coil linked with it.	94		
	Recognise and use the expression $\mathcal{E} = -d(N\Phi)/dt$.	95		
	Explain how $\mathcal{E} = -d(N\Phi)/dt$ is a consequence of Faraday's and Lenz's laws.	95		

ResultsPlus
Build Better Answers

The voltage across a capacitor varies with the charge on the capacitor as in the graph.

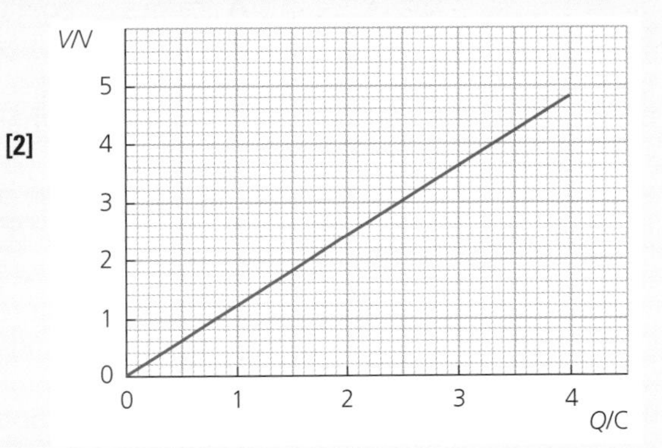

a i Calculate a value for the capacitance of the capacitor. **[2]**

Student answer	Examiner comments
$C = \frac{Q}{V} = \frac{4.0}{4.4} = 0.91$ F	The student has read the voltage incorrectly from the graph, so the answer is wrong. However 1 mark was gained for a substitution into the correct formula.

ii Use the graph to derive the expression $W = \frac{1}{2}QV$ for the energy stored by a capacitor. **[3]**

Examiner tip

Check the specification for formulas that you need to be able to derive. Learn the derivations ready for the exam, as derivations can be tricky to reproduce if you haven't practised them in advance.

Student answer	Examiner comments
Energy stored = area under graph = $\frac{1}{2}Q \times V$	The answer scores 1 mark for stating the relationship between energy stored and area, but there is no justification given. The student should have referred to work done $W = Q \times V_{av}$ and explained that they have given the area of a triangle.

iii Calculate the energy stored by the capacitor when the voltage across it is 4.0 V. **[2]**

Student answer	Examiner comments
$W = \frac{1}{2}QV = 0.5 \times 4.8 \times 4.0 = 9.6$ J	The answer scores no marks, as once again the graph has been misread. This time Q and V have been mixed up, and the voltage for a *charge* of 4.0 C has been read off. Check your answers again after you have written them down to avoid this common kind of error.

b The graph on the right shows how the charge on the capacitor varies with time when the capacitor is used to light an LED.

i Explain why the output power from the LED decreases with time. **[2]**

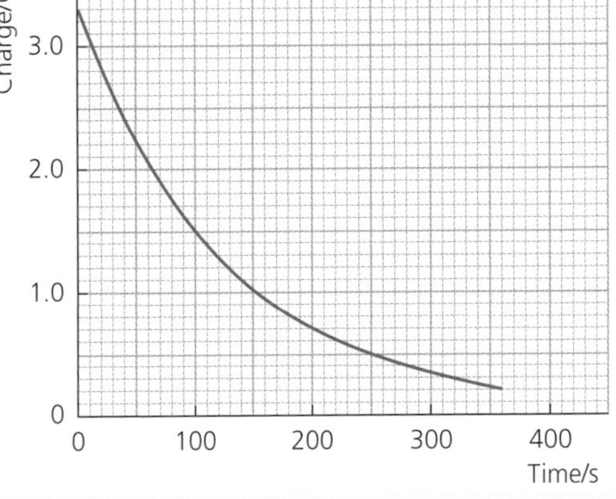

Examiner tip

In any explanation that you give, check that each step follows on logically. Don't assume that the examiner will fill in the missing bits for you.

Student answer	Examiner comments
As current flows the voltage across the capacitor decreases. Hence the power decreases ($P = IV$).	The answer scores 1 mark, as a correct formula for power has been given. However, the link between charge on the capacitor and current in the circuit is not stated, and so it is unclear from the graph given why V should decrease.

ii Use the graph to determine the time constant for the capacitor–LED circuit. **[1]**

Edexcel June 2007 Unit 8 PSA 4

Student answer	Examiner comments
$Q = 3.2$ C when $t = 0$. $Q = 1.6$ C when $t = 90$ s. So the time constant is 90 s.	Time constant has been mistaken for half life. The time constant is the time taken for the charge to fall to 37% of its initial value, and so is about 130 s.

Practice exam questions

1 An oil drop is suspended in equilibrium between two parallel plates. There is a uniform electric field acting downwards between the plates. The oil drop

A carries a negative charge
B carries a positive charge
C is uncharged
D may carry either type of charge **[1]**

2 When the voltage placed across a capacitor is doubled,

A the charge and energy stored doubles
B the charge stored doubles and the energy quadruples
C the energy stored doubles and the charge quadruples
D the energy and charge stored quadruples **[1]**

3 A capacitor of capacitance 2.2 μF is charged up through a resistor of resistance 47 MΩ. The time taken for the capacitor to completely discharge is approximately

A 20 s **B** 100 s **C** 110 s **D** 520 s **[1]**

4 An aeroplane has a wingspan of 60 m and is moving horizontally at a speed of 110 m s^{-1}. The Earth's magnetic field has a vertical component of 50 μT and a horizontal component of 20 μT. The magnitude of the emf generated between the wing tips is

A 0 V **B** 0.13 V **C** 0.33 V **D** 0.36 V **[1]**

5 An oil drop weighs 5.8×10^{-15} N. It is suspended in a uniform field produced by a potential difference of 363 V between two parallel plates that are 6.0 cm apart.

a Calculate the charge of the drop. **[4]**
b How many excess electrons does it carry? **[1]**

6 Electrons are accelerated in an electron gun of a cathode ray tube.

a Calculate the velocity of electrons emerging from the gun if the anode voltage is 350 V. **[3]**

+75 V

e$^-$

0 V

The electron beam passes through a pair of parallel plates separated by 2.5 cm across which there is a potential difference of 75 V as shown oppposite.

b Copy and complete the diagram to show **i** the electric field and **ii** the path of the electron beam. **[2]**
c Calculate the strength of the electric field between the plates. **[2]**
d A magnetic field is applied in the region between the plates so that there is no deflection of the electrons as they pass through the plates. Give the direction in which must the magnetic field be applied, and justify your answer. **[2]**
e Calculate the strength of the magnetic field required to give no deflection of the beam. **[3]**

7 A 2.2 μF capacitor is charged to a potential difference of 15 V, and a 3.3 μF capacitor is charged to a potential difference of 30 V.

a Calculate

i the charge on each capacitor, **[2]**
ii the energy stored by each capacitor. **[2]**

b The capacitors are then joined together as in the circuit shown.
When the switch S is closed, charge re-distributes between the capacitors.
Explain why the final voltage across each capacitor is the same. **[1]**

S
2.2 μF
3.3 μF
+
–
+
–

c Show that, once charge has re-distributed, the ratio of charge on each capacitor is equal to the ratio of the two capacitances. Hence calculate the charge on each capacitor and the final potential difference. **[4]**

d Calculate the total energy stored by the capacitors and suggest why this is less than the total energy that would be stored by the capacitors individually. **[3]**

8 A student is designing an alarm circuit. She wants a time interval between closing a switch and the circuit becoming active. She decides to charge a 2200 μF capacitor through a resistor from a 5 V supply. The potential difference across the capacitor must rise to 4.3 V for the alarm circuit to become active.

Calculate the size of the resistor that she should connect in the circuit to achieve a 1 minute delay. **[3]**

9 A teacher demonstrates electromagnetic induction by dropping a bar magnet from rest through the centre of a coil of wire. A data logger records the emf in the coil.

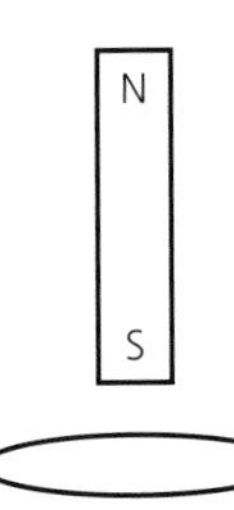

a i Why is a data logger necessary for this demonstration? **[2]**

ii State the magnetic polarity on the upper side of the coil due to the induced current as the magnet approaches it. **[1]**

The graph shows the emf induced in the coil as the magnet falls.

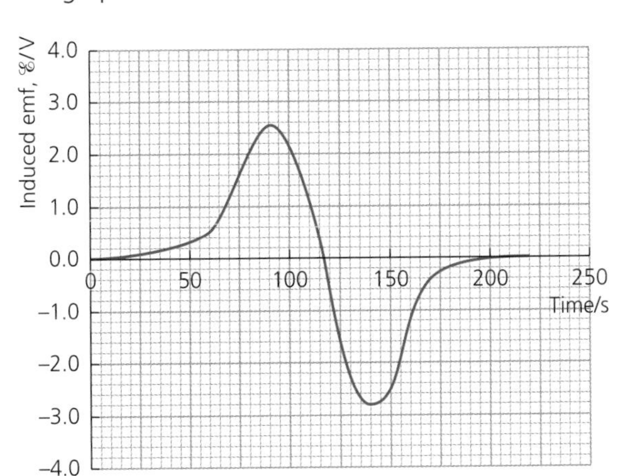

b Explain why the maximum negative induced voltage induced is greater than the maximum positive induced voltage. **[2]**

c A student comments that the areas between the graph above and below the time axis look similar. Suggest why you would expect these areas to be identical. **[2]**

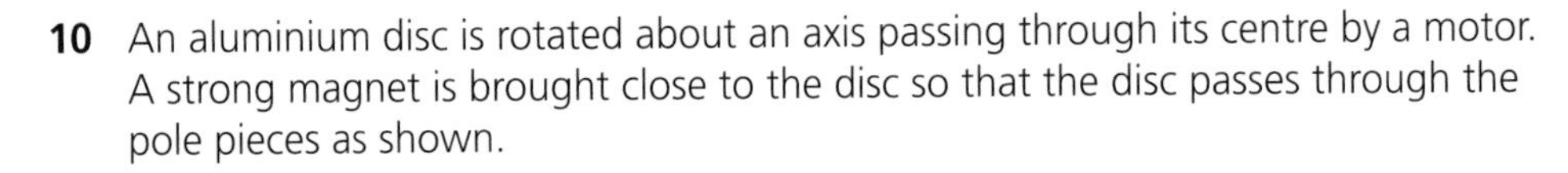

10 An aluminium disc is rotated about an axis passing through its centre by a motor. A strong magnet is brought close to the disc so that the disc passes through the pole pieces as shown.

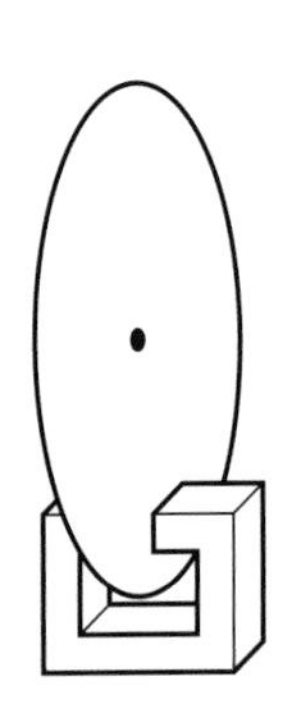

When the magnet is introduced, the rate of rotation of the disc decreases. Use the laws of Faraday and Lenz to explain why this happens. **[5]**

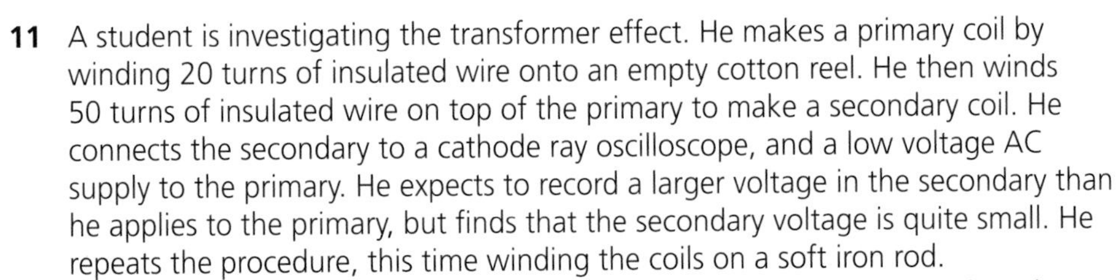

11 A student is investigating the transformer effect. He makes a primary coil by winding 20 turns of insulated wire onto an empty cotton reel. He then winds 50 turns of insulated wire on top of the primary to make a secondary coil. He connects the secondary to a cathode ray oscilloscope, and a low voltage AC supply to the primary. He expects to record a larger voltage in the secondary than he applies to the primary, but finds that the secondary voltage is quite small. He repeats the procedure, this time winding the coils on a soft iron rod.

a Explain why you would expect the secondary voltage to be larger than the primary voltage this time. **[2]**

b The AC supply is replaced by a steady DC supply with the same output voltage as the AC supply.

i Explain why there is no voltage produced in the secondary. **[2]**

ii The student notices that the primary coil quickly gets very warm with the DC supply. Suggest why this was not so noticeable with the AC supply. **[3]**

The nuclear atom

Nuclear structure

The nucleus of an atom consists of protons and neutrons, collectively called **nucleons**. The **proton number** (or atomic number), Z, determines the element. Atoms with the same number of protons but different numbers of neutrons are called **isotopes**. Isotopes have different **nucleon numbers** (or mass numbers), A.

We represent a nucleus in symbols as ${}^{A}_{Z}X$ where X is its chemical symbol. A helium nucleus consists of 2 protons and 2 neutrons and so we write ${}^{4}_{2}He$ for the normal isotope of helium.

The balance of neutrons and protons in a nucleus is crucial to its stability. An unstable nucleus may emit an **α-particle** (equivalent to a helium nucleus) or a **β^--particle** (a fast electron from the nucleus). This decay process changes the nuclear configuration to achieve greater stability. **γ-radiation** may also be emitted by some nuclides, although no change in the nuclear configuration occurs. An example of α-decay:

$$ {}^{220}_{86}Rn \rightarrow {}^{216}_{84}Po + {}^{4}_{2}He + \text{energy} $$

The nucleon number stays the same before and after the decay (here 220 = 216 + 4), as does the proton number (here 86 = 84 + 2). This is also true for β-decay. In all nuclear decays, **nucleon and proton numbers are conserved**.

Worked Example

${}^{131}_{53}I$ is an unstable isotope of iodine that emits β^--particles. Write an equation to show the decay of this isotope into xenon, Xe.

$$ {}^{131}_{53}I \rightarrow {}^{131}_{54}Xe + {}^{0}_{-1}\beta + {}^{0}_{0}\bar{\nu} $$

Alpha scattering

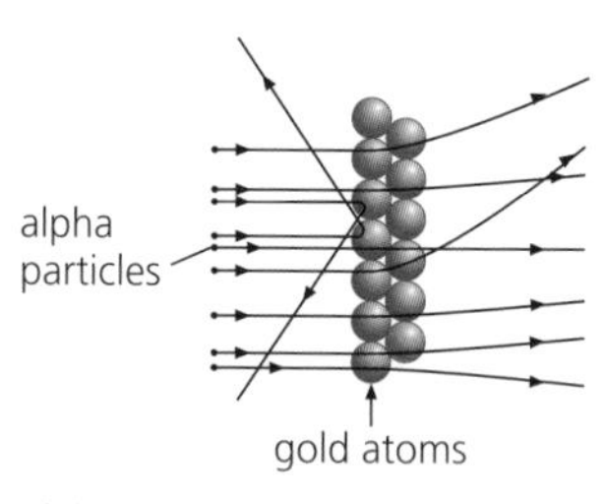

Alpha particle scattering

Evidence for the nuclear model of the atom was obtained from an experiment in which α-particles were fired through thin gold foils. The vast majority of the α-particles experienced very little or no deflection as they passed through the foil. However, a tiny fraction of the α-particles were deflected through large angles – some even rebounded backwards.

The **nuclear model** explains this by saying that the rebounding atoms were deflected by a tiny, dense, positively charged nucleus.

Electron beams

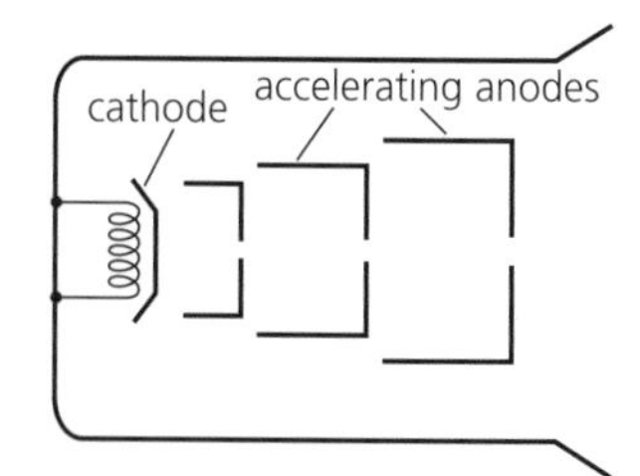

An electron gun

Beams of particles are important in investigating sub-atomic structure. In cathode ray tubes an 'electron gun' is used to produce a beam of fast electrons. A filament heats a metal cathode which releases electrons via **thermionic emission**. The electrons are accelerated to high velocities through a large potential difference.

Worked Example

Electrons are accelerated through a potential difference of 1000 V in the electron gun of a cathode ray tube. Show that the speed of the electrons as they leave the gun is about $2 \times 10^7\,\text{m s}^{-1}$.

Gain in kinetic energy of electrons = work done by electric field

$$\frac{1}{2}mv^2 = eV \quad (u = 0)$$

$$v^2 = \frac{2eV}{m} = \frac{2 \times 1.60 \times 10^{-19}\,\text{C} \times 1000\,\text{V}}{9.11 \times 10^{-31}\,\text{kg}}$$

$$v = \sqrt{\frac{1.60 \times 10^{-16}}{4.55 \times 10^{-31}}}$$

$$v = 1.88 \times 10^{7}\ \text{m s}^{-1}$$

Quantities such as the charge and mass of an electron may not be given in the question – check the data sheet at the end of the exam paper!

Electrons as waves

Electrons do not just behave as particles – they also have wave properties. **De Broglie's wave equation** relates the wavelength, λ, of a particle to its momentum, p.

$$\lambda = \frac{h}{p}$$

where h is the **Planck constant**, equal to 6.63×10^{-34} J s.

Fast electrons or other particles may have a wavelength similar in size to nuclear matter, which means its structure can be investigated in scattering experiments. The higher the particle energy, the shorter the wavelength, hence the greater the detail that can be resolved. So high-energy particle beams are required for fine structure to be investigated.

Worked Example

Electrons are accelerated to a speed of 10% of the speed of light in the electron gun of a cathode ray tube. Show that the de Broglie wavelength of the electrons is similar to that of the spacing between atoms (which is about 10^{-10} m).

Speed of light = 3.00×10^8 m s^{-1}, so electron speed

$$v = 0.1 \times 3.00 \times 10^8\ \text{m s}^{-1} = 3.00 \times 10^7\ \text{m s}^{-1}$$

Substitute into $\lambda = \frac{h}{mv}$:

$$\lambda = \frac{6.63 \times 10^{-34}\ \text{J s}}{9.11 \times 10^{-31}\ \text{kg} \times 3.00 \times 10^7\ \text{m s}^{-1}} = 2.4 \times 10^{-11}\ \text{m}$$, which is similar to atomic spacing.

Thinking Task

Electrons are accelerated in an electron gun through a potential difference of 3.5 kV. Show that the velocity of the electrons as they emerge from the gun is about 4×10^7 m s^{-1}. Hence calculate the de Broglie wavelength of these electrons. Why would these electrons be unsuitable for investigating nuclear structure?

Quick Questions

Q1 The diagram on the right shows an α-particle being deflected as it approaches a gold nucleus. Copy it, and mark the direction of the force on the α-particle at the points A, B, and C.

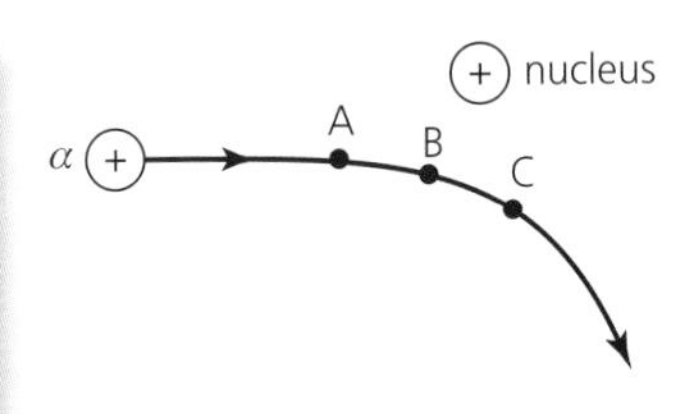

Q2 **a** Complete the equation below, which shows the fission of uranium-235, by working out the numbers w, x, y and z.

$$^{236}_{92}\text{U} \rightarrow {}^{144}_{56}\text{Ba} + {}^{89}_{w}\text{Kr} + x\,{}^{z}_{y}\text{n}$$

b Barium-144 is unstable, and decays to lanthanum, La, via β^--decay. Write an equation to represent this decay.

Q3 Calculate the de Broglie wavelength of electrons fired from an electron gun with a speed of 1.0×10^7 m s^{-1}.

High-energy collisions

Producing new particles

New particles can be produced when unstable particles decay or during high-energy particle interactions in particle accelerators. When matter is made to collide with anti-matter, **annihilation** results. The energy released is used to form new particles. According to Einstein's **mass–energy equation**:

$$\Delta E = c^2 \Delta m$$

where c is the speed of light and ΔE is the energy equivalent of a mass Δm.

The greater the energy of the colliding particles the greater the energy available to produce new particles. The kinetic energy of the particles before the collision is also available for the formation of new particles.

Worked Example

The electron mass is 9.11×10^{-31} kg. Calculate the energy released when an electron and a positron annihilate. A positron has the same mass as an electron.

$$\Delta E = (3.00 \times 10^8 \, \text{m s}^{-1})^2 \times 2 \times 9.11 \times 10^{-31} \, \text{kg} = 1.60 \times 10^{-13} \, \text{J}$$

Check carefully which units the question requires – sometimes questions may give energies in eV but require an answer in joules (and vice versa).

Energy units

Particle physicists usually refer to particle energies in **electron-volts** (eV) or their multiples (MeV and GeV). By definition, the electron-volt is the energy transformed when an electron moves through a potential difference of 1 V. Since $E = QV$ and $Q = 1.6 \times 10^{-19}$ C for an electron, then 1 eV = 1.6×10^{-19} J.

Linear accelerators

In a **linear accelerator**, or **linac**, an alternating voltage is used to create an alternating electric field that accelerates charged particles.

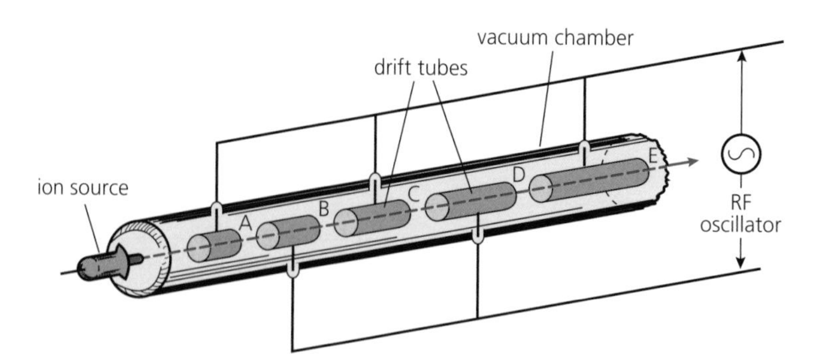

The construction of a linear accelerator

Charged particles travel at constant speed inside the drift tubes, but are accelerated each time they cross a gap between tubes. The drift tubes increase in length along the linac to keep the time taken to travel to the next acceleration point constant. This matches with the frequency of the alternating signal being applied to the tubes.

Worked Example

Electrons are accelerated by an alternating voltage of 5 kV applied to the electrodes of a linac. If the electrons cross 3000 gaps before reaching the end of the linac, calculate the energy of the electrons emerging from the linac in MeV.

Apply $E = QV$ for energy gained when particle crosses *each gap*:

$$\text{total } E = 3000 \times 1.6 \times 10^{-19}\,\text{C} \times 5000\,\text{V} = 2.4 \times 10^{-12}\,\text{J}$$

Use 1 MeV = 1.6×10^{-13} J, giving

$$\text{total } E = \frac{2.4 \times 10^{-12}\,\text{J}}{1.6 \times 10^{-13}\,\text{J MeV}^{-1}} = 15\,\text{MeV}$$

These multiples of eV are all commonly used in particle physics:

keV (1×10^3 eV)
MeV (1×10^6 eV)
GeV (1×10^9 eV).

Worked Example

When two particle beams meet head on, more energy is available than when the particle beam is directed at a fixed target. Why is this?

In a target experiment, the momentum of any particles produced has to equal the momentum of the beam particle. So the particles produced must have lots of kinetic energy. Since total energy is also conserved, this is at the expense of the rest mass energy. In a colliding beam experiment the initial momentum is zero, so *all* the initial energy is available for creating the mass of new particles.

More non-SI units

Particle masses are all tiny fractions of a kilogram, so particle physicists use more conveniently sized units. The **atomic mass unit**, u, is defined as being 1/12 of the mass of an atom of the carbon isotope ^{12}C.

$$1\,\text{u} = 1.66 \times 10^{-27}\,\text{kg}$$

The unit **eV/c^2** is also used. This arises from the Einstein mass–energy equation $\Delta E = c^2 \Delta m$. We use the energy equivalent of the mass (in MeV or GeV) and indicate that it is a mass by writing MeV/c^2 or GeV/c^2.

Worked Example

An α-particle has a mass of 4.032 u. Calculate its mass in MeV/c^2.

$$\text{mass of } \alpha = 4.032 \times 1.66 \times 10^{-27}\,\text{kg} = 6.69 \times 10^{-27}\,\text{kg}$$

$$\text{energy equivalent of } \alpha = 6.69 \times 10^{-27}\,\text{kg} \times (3.00 \times 10^8\,\text{m s}^{-1})^2 = 6.02 \times 10^{-10}\,\text{J}$$

$$= \frac{6.02 \times 10^{-10}\,\text{J}}{1.60 \times 10^{-19}\,\text{J eV}^{-1}} = 3.76 \times 10^9\,\text{eV} = 3760\,\text{MeV}$$

$$\therefore \text{mass of } \alpha = 3760\,\text{MeV}/c^2$$

Quick Questions

Q1 The length of the drift tubes increases along a linac. Explain what would happen to the energy of the charged particles as they progressed if the drift tubes were all the same length.

Q2 Protons are accelerated to an energy of 10 GeV and the beam is made to collide head-on with a beam of anti-protons, also with an energy of 10 GeV. How much energy is available from each collision to produce new particles? Take the proton mass as 940 MeV/c^2.

Thinking Task

Rutherford bombarded nitrogen-14 with α-particles, producing oxygen-17 and a high-energy proton:

$$^{14}_{7}\text{N} + ^{4}_{2}\alpha + ^{17}_{8}\text{O} + ^{1}_{1}\text{p}$$

The α-particle energy used by Rutherford was 7.70 MeV. Using the data in the table, calculate the energy to be shared after the interaction, and explain why the proton cannot take all of the available energy.

Nuclide	Mass/u
$^{1}_{1}$H	1.007 825
$^{4}_{2}$He	4.002 603
$^{14}_{7}$N	14.003 074
$^{17}_{8}$O	16.999 132

Particle accelerators

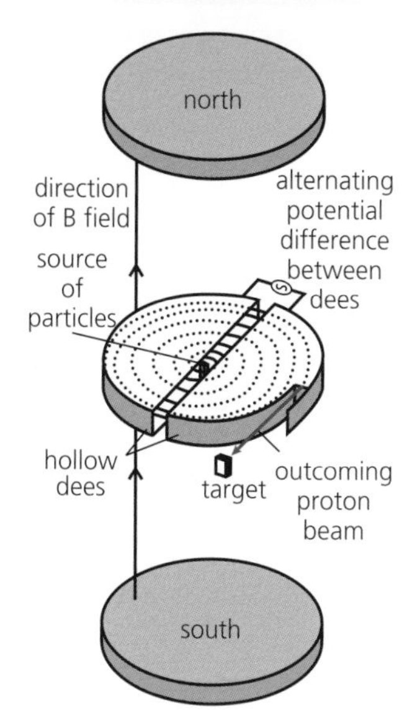

The construction of a cyclotron

The cyclotron

Linear accelerators can produce very high-energy particles but they need to be very long. A **cyclotron** uses a magnetic field to bend the particle beam into a circular path.

Charged particles are injected near the centre of the cyclotron. They are accelerated by an electric field as they cross the gap between the 'dees' (see diagram), and are deflected into a circular path by a magnetic field as they move inside the dees. The electric field is zero within the dees.

When the particles are in the magnetic field they experience a force according to Fleming's left hand rule (see page 26). This force keeps the particles in a circular path, but does not change their speed. The motion of a charged particle in a magnetic field is given by

$$r = \frac{p}{BQ}$$

where r is the radius of a curved path, p is momentum ($p = mv$), B is the magnetic field strength, and Q is the charge.

As the particles are accelerating, the radius of their path increases and they spiral outwards inside the dees. The time to move around the dees is independent of speed, as long as the mass of the particles does not change. Hence the alternating electric field is able to accelerate the particles each time they cross the gap between the dees. However, once particles approach the speed of light relativistic effects become important and the mass of the particles increases. The frequency of the particles moving around the cyclotron no longer matches that of the applied potential difference, and so the particles do not continue to gain energy as they move around the cyclotron.

This limitation means that the maximum energy that particles can be given in a cyclotron is about 20 MeV. This is of limited use to high-energy particle physicists, but is sufficient for manufacturing isotopes for medical or industrial use.

Worked Example

A beam of electrons travelling at 15% of the speed of light enters a uniform magnetic field of strength 0.2 T. What is the radius of the circular path that they are deflected into?

$$v = 0.15 \times 3.0 \times 10^{8}\,\text{m s}^{-1} = 4.5 \times 10^{7}\,\text{m s}^{-1}$$

$$r = \frac{mv}{BQ}$$

$$= \frac{9.1 \times 10^{-31}\,\text{kg} \times 4.5 \times 10^{7}\,\text{m s}^{-1}}{0.2\,\text{T} \times 1.6 \times 10^{-19}\,\text{C}} = 1.3 \times 10^{-3}\,\text{m}$$

Worked Example

Electrons in a cyclotron emit radio waves at a frequency equal to their frequency of rotation. What is the magnetic field strength if the radio waves are at a frequency of 11.5 MHz?

Calculate the time taken for the electrons to make one revolution, then use $f = \frac{1}{T}$.

$$T = \frac{2\pi r}{v} \quad f = \frac{1}{T} = \frac{v}{2\pi r}$$

Substitute $r = \frac{mv}{BQ}$:

$$f = \frac{vBQ}{2\pi mv} = \frac{BQ}{2\pi m}$$

Rearrange and substitute the value of f to calculate B:

$$B = \frac{2\pi mf}{Q}$$

$$= \frac{2\pi \times 9.11 \times 10^{-31}\,\text{kg} \times 11.5 \times 10^{6}\,\text{s}^{-1}}{1.60 \times 10^{-19}\,\text{C}}$$

$$= 4.1 \times 10^{-4}\,\text{T}$$

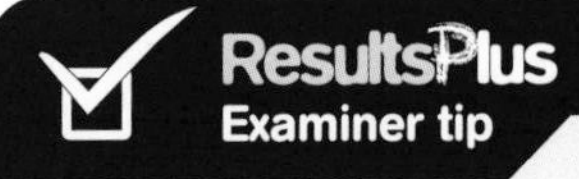

The question seems open ended, but the information given points to use of a formula for cyclotron frequency as a starting point. Look for clues in the question when the way forward isn't clear.

Quick Questions

Q1 A cyclotron with dees of diameter 1.6 m is used with a magnetic field of strength 0.75 T. Show that the maximum speed of protons produced by this accelerator is about $6 \times 10^{7}\,\text{m s}^{-1}$.

Q2 Show that the maximum energy that can be given to an electron in a cyclotron of given diameter depends upon the square of the magnetic field strength used to deflect the electrons.

Q3 A β^{-}-particle is emitted with a speed of $3 \times 10^{7}\,\text{m s}^{-1}$ from a radioactive source in a cloud chamber. A magnetic field of strength 4.5 mT is applied at right angles to the initial direction of the β^{-}-particle. Calculate the radius of the path followed by the β^{-}-particle. Make a sketch to show the β^{-}-particle path and the magnetic field direction.

Particle theory

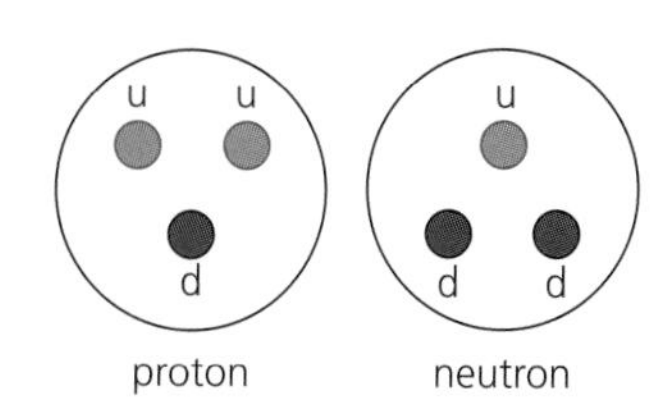

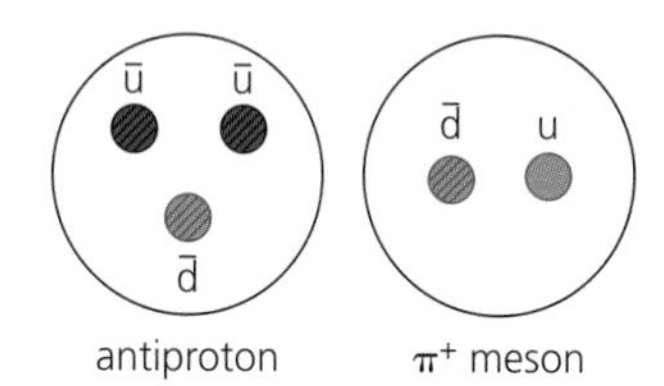

The quark arrangement of some hadrons

	Quarks		Leptons	
	u	d	e^-	ν_e
	c	s	μ^-	ν_μ
	t	b	τ^-	ν_τ
Charge / e	$\frac{2}{3}$	$-\frac{1}{3}$	-1	0

The standard model

Our current ideas about fundamental particles are summarised by the **standard model** of particle theory. In this model there are two types of fundamental particle – **quarks** and **leptons**.

Quarks are **strongly interacting particles**, and it is believed they do not exist singly. They occur in two possible combinations:

- in quark–anti-quark pairs, called **mesons**
- in quark triplets, called **baryons**.

Collectively mesons and baryons are referred to as hadrons. The proton and the neutron are familiar examples of baryons.

Leptons are **weakly interacting particles**. Leptons can occur singly; the electron is a familiar example of a lepton.

In the standard model there are six quarks and six leptons, plus their anti-particle equivalents. The **up** (u) and **down** (d) quarks were the first to be discovered, followed by the **strange** (s) and **charmed** (c) quarks. On symmetry grounds, a third generation of quarks called **top** (t) and **bottom** (b) was suggested and subsequently discovered. The table shows the three generations of fundamental particle in the standard model.

Worked Example

The K^- meson is a light particle that has been observed to decay via the weak interaction into a π meson and other particles:

$$K^- \rightarrow \pi^0 + X + \bar{\upsilon}_\mu$$

a The K^- contains a strange quark. Given that the charge on a strange quark is $-\frac{1}{3}$, deduce the composition of the K^- meson.

b What is the charge of particle X? Justify your answer.

---

a K^- is a meson, so it is a quark–anti-quark pair. If it contains a strange quark of charge $-\frac{1}{3}$, it must contain an anti-quark of charge $-\frac{2}{3}$, so that the overall charge is -1. The up quark has a charge of $\frac{2}{3}$, so there must be an anti-up quark in the combination. Hence the K^- meson has a composition $\bar{u}s$.

b Particle X must have a charge -1, because neither the pion nor the anti-neutrino has any charge, and the negative charge of the K^- meson must be conserved.

Detecting new particles

The nature of new particles formed in high-energy collisions is investigated by analysing their tracks in detectors. The particles ionise (remove electrons from) atoms in detectors. In a bubble chamber, the ionised atoms form sites where bubbles can form so that the tracks can be made visible. In spark chambers or Geiger counters, the released electrons produce a current that can be detected.

Electric and magnetic fields are used to cause deflections in the particles' tracks. The deflection in an electric field gives information on the charge of the particle, and the deflection in a magnetic field gives information on the momentum of the particle (remember that $p = BQr$).

When new particles are formed, the fundamental conservation laws must be upheld. In particular, charge, energy, and momentum must be the same before and after the collision.

Worked Example

A particle entering from the bottom of the diagram decays into three particles at A. One of the particles leaves no track, but subsequently decays into two particles at B.

a Why does the particle that decays at B leave no track as it travels from A to B?

b There is a magnetic field directed into the plane of the paper. What can you conclude about the charge and momentum of particles C and D? Justify your conclusions.

---

a The particle is uncharged, and so does not cause ionisation inside the detector.

b Particles C and D must have equal and opposite charge, because charge must be conserved, and because the tracks curve in opposite directions in the magnetic field. Applying Fleming's left hand rule, particle C must be negatively charged.

Particles C and D must have the same momentum, as $r = \frac{p}{BQ}$, and both tracks have the same radius of curvature.

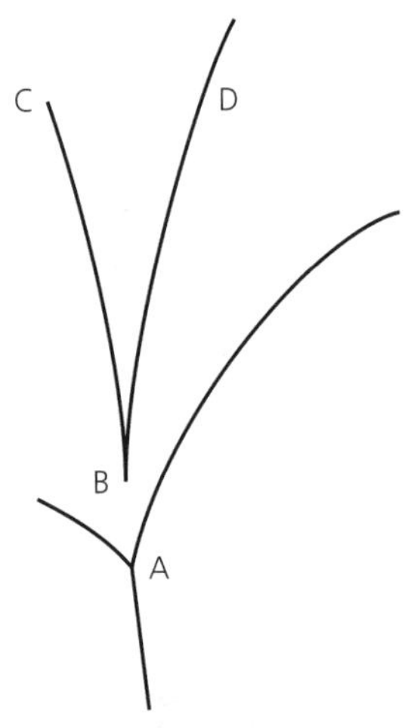

Drawing of particle tracks in a bubble chamber

Quick Questions

Q1 In β^+-decay a proton within a nucleus decays into a neutron, and a positron and a neutrino are emitted from the nucleus. Write an equation to represent this process, and state how the proton's quark structure changes in the decay.

Q2 The Λ^0 is a baryon of mass 1.116 GeV/c^2 that provided the first observational evidence for the strange quark. It decays via the weak interaction to form a neutron and a pion (π meson).

a State what is meant by the term *baryon*. Write an equation to represent the decay.

b If the Λ^0 is at rest when it decays, how much kinetic energy is available to be shared between the neutron and pion? The masses of the neutron and pion are 940 MeV/c^2 and 135 MeV/c^2 respectively.

Thinking Task

The B meson is a heavy particle containing a $\bar{b}$ antiquark and either a u or a d quark.

a Write down the quark structure of a B^+-particle, justifying your choice.

b The B^+-particle may be produced along with the B^--particle in high-energy collisions between protons and anti-protons. If the B meson has a mass of 5.28 GeV/c^2, calculate the minimum kinetic energy of the proton necessary for a B^+, B^- pair to be produced from a proton–anti-proton head-on collision. The proton has a mass of 938 MeV/c^2.

Section 3: Particle physics checklist

By the end of this section you should be able to:

Revision spread	Checkpoints	Spec. point	Revised	Practice exam questions
The nuclear atom	Use the terms nucleon number (mass number) and proton number (atomic number).	96	☐	☐
	Describe how large-angle alpha particle scattering gives evidence for a nuclear atom.	97	☐	☐
	Recall that electrons are released in the process of thermionic emission and explain how they can be accelerated by electric and magnetic fields.	98	☐	☐
	Explain why high energies are required to break particles into their constituents and to investigate fine structure.	102	☐	☐
	Write and interpret equations using standard nuclear notation.	107	☐	☐
	Use de Broglie's wave equation $\lambda = \frac{h}{p}$.	108	☐	☐
High-energy collisions	Explain the role of electric fields in linear particle accelerators.	99	☐	☐
	Recognise and use the expression $r = \frac{p}{BQ}$ for a charged particle in a magnetic field.	100	☐	☐
	Recognise and use the expression $\Delta E = c^2 \Delta m$ in situations involving the creation and annihilation of matter and anti-matter particles.	103	☐	☐
	Use the non-SI units MeV and GeV (energy) and MeV/c^2, GeV/c^2 (mass) and atomic mass unit u, and convert between these and SI units.	104	☐	☐
Particle accelerators	Explain the role of electric and magnetic fields in cyclotron particle accelerators.	99	☐	☐
	Be aware of relativistic effects and that these need to be taken into account at speeds near that of light (use of relativistic equations not required).	105	☐	☐
Particle theory	Explain the role of electric and magnetic fields in particle detectors.	99	☐	☐
	Recall and use the fact that charge, energy and momentum are always conserved in interactions between particles and hence interpret records of particle tracks.	101	☐	☐
	Recall that in the standard quark–lepton model each particle has a corresponding anti-particle, that baryons (e.g. neutrons and protons) are made from three quarks, and mesons (e.g. pions) from a quark and an anti-quark, and that the symmetry of the model predicted the top and bottom quark.	106	☐	☐
	Write and interpret equations using standard symbols (e.g. π^+, e^-).	107	☐	☐

ResultsPlus
Build Better Answers

Results of high-speed collisions between electrons and protons have suggested that protons have a bulging rather than a spherical shape. These results have been explained by modelling a nucleon as 'a relativistic system of three bound quarks surrounded by a cloud of pions'.

Protons and neutrons are the two types of nucleon and both consist of up and down quarks.

Nucleon	Quark composition
proton	uud
neutron	udd

Quark	Charge
up	$\frac{+2e}{3}$
down	$\frac{-e}{3}$

a Use the information in the tables to deduce the charge of the proton and the neutron. **[2]**

Student answer	Examiner comments
Proton = uud so charge = $\frac{2e}{3} + \frac{2e}{3} - \frac{e}{3} = +e$ Neutron = udd so charge = $\frac{2e}{3} - \frac{e}{3} - \frac{e}{3} = 0$	The student has explained how they arrived at their answer, so the answer scores full marks.

b Protons, neutrons and pions are all hadrons. There are two types of hadron, with different quark combinations.

i Complete the table on the right to name the two types of hadron. **[2]**

Particles	Hadron type
Proton, neutron	
Pion	

Student answer	Examiner comments
Boson, meson	Although the pion is an example of a meson, protons and neutrons are **baryons**, so only 1 mark is gained for this answer.

ii State the differences in quark composition between these two types of hadron. **[2]**

Student answer	Examiner comments
One has three quarks, the other has two quarks.	This scores no marks, as the student has not said which type of hadron has three quarks. Also, saying that mesons consist of two quarks lacks precision, because they are quark–anti-quark pairs.

c Explain why high-speed particles are used to examine the internal structure of other particles. **[4]**

Examiner tip

In an explanation that involves a number of points it is often a good idea to use a bulleted list for your answer. Make sure you have as many points as there are marks available.

Student answer	Examiner comments
• high speed allows particles to overcome repulsive forces • high speed also means short wavelength $\left(\lambda = \frac{h}{p}\right)$ • for diffraction we need λ to be very small.	The first point is vague; it would have been better to make the connection between high speed and high energy. The second point is worth 2 marks as the relationship between speed and wavelength is correct, and the de Broglie relationship is also quoted. The third point could be improved by saying that the wavelength must be about the size of the spacing between the particles. Nonetheless this answer is worth 4 marks, as the mark scheme allowed for a variety of answers.

d The model mentions a 'relativistic system'. State the conditions needed for relativistic effects to be significant. **[1]**

Edexcel June 2008 Unit PSA4

Student answer	Examiner comments
The speed must be near the speed of light.	This is correct and scores 1 mark.

Practice exam questions

1 Which of the following is a correct conclusion from Rutherford's α-scattering experiments?

A α-particles have wave-like properties.
B Gold is a very dense metal.
C The nucleus contains neutrons.
D Thompson's plum pudding model is incorrect. **[1]**

2 Which one of the following is *not* a hadron?

A a kaon
B a neutrino
C a neutron
D a pion **[1]**

3 Which one of the following particles would *not* leave a track in a bubble chamber?

A electron
B neutron
C positron
D proton **[1]**

4 High-energy particles are needed to investigate fine structure because such particles have

A small momentum and small wavelength
B small momentum and large wavelength
C large momentum and small wavelength
D large momentum and large wavelength **[1]**

5 The thorium series is a naturally occurring radioactive decay chain. Thorium-232 is the beginning of this chain and the final stable isotope is lead-208. Lead is the largest naturally occurring stable atom, with 82 protons in the nucleus.

a Thorium-232 is an α-emitter. Complete the decay equation. **[2]**

$$^{232}_{90}\text{Th} \rightarrow {}^{?}_{?}\text{Ra} + {}^{?}_{?}\alpha$$

b Lead-208 forms from the β^--decay of thallium-208. Complete the decay equation. **[3]**

$$^{208}_{?}\text{Tl} \rightarrow {}^{208}_{?}\text{Pb} + {}^{?}_{?}\beta^- + {}^{?}_{?}\overline{\upsilon}_e$$

c In β^--decay a neutron decays into a proton in the nucleus. How does the neutron's quark structure change in this decay? In the decay the particle $\overline{\upsilon}_e$ is also produced. What sort of fundamental particle is this? **[2]**

d How many α-particles and how many β^--particles are emitted when a thorium-232 nucleus decays to lead-208? **[3]**

6 A cyclotron is used to produce a beam of high-energy protons for the treatment of eye tumours. A high-frequency alternating pd of 42 kV applied between the dees is used to give energy to the protons. A uniform magnetic field of strength 1.65 T applied perpendicular to the plane of the dees deflects the protons into a circular path.

a Explain why the protons are deflected into a circular path by the magnetic field. State with a reason the direction in which the magnetic field must be applied to deflect the protons as in the diagram. **[3]**

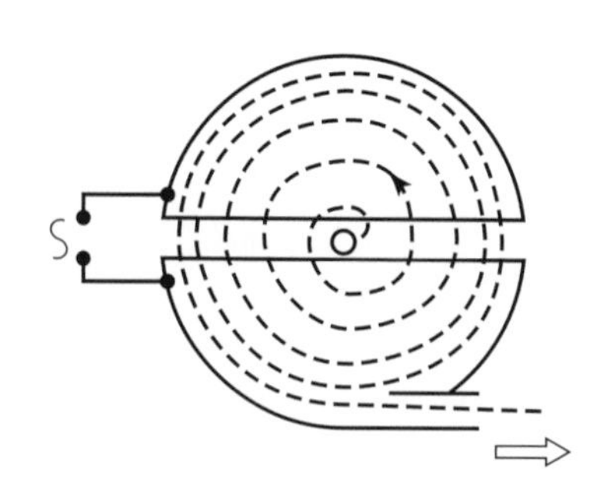

b Show that the gain in energy of a proton accelerated through a potential difference of 42 kV is about 7×10^{-15} J. **[2]**

c Calculate the frequency of the alternating supply that must be applied to the gap between the dees in order to maintain the energy increase calculated in **b** each time the protons cross the gap between the dees. **[3]**

d The protons cross the gap between the dees 1500 times before they emerge from the cyclotron. Calculate the energy acquired by the protons, and hence the radius of the circle in which they will be moving just before they emerge. **[3]**

e Give a reason why there is a maximum energy that can be given to the protons. **[2]**

7 The diagram shows tracks of particles created in a bubble chamber when an anti-proton annihilates with a proton. A magnetic field was directed into the chamber in the plane of the diagram.

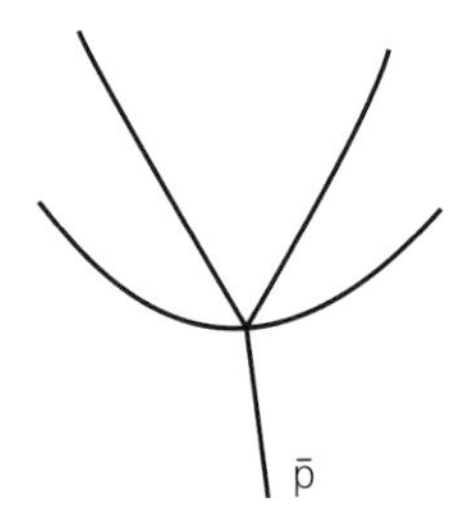

The interaction is represented by the following equation:

$$p + \bar{p} = \pi^+ + \pi^+ + \pi^0 + \pi^- + \pi^-$$

a Explain why only five tracks are visible in the bubble chamber photograph. **[1]**

b Copy the diagram and label the two π^+ particles. How do you know that the tracks that you have labelled are made by positively charged particles? **[2]**

c The π^+ particles have different energies. Identify the particle with the higher energy, justifying your choice. **[2]**

d Protons and pions are examples of hadrons. Give one similarity and one difference between protons and pions. **[2]**

e The anti-proton had an energy of 1.0 GeV. If the proton was initially at rest, how much energy was available for kinetic energy of the pions? The proton mass is 940 MeV/c^2, the mass of a neutral pion is 135 MeV/c^2 and that of a charged pion is 140 MeV/c^2. **[1]**

8 An electron diffraction tube contains an electron gun to produce a high-energy beam of electrons. The electrons pass through a thin graphite section and are detected on a fluorescent screen. Diffraction rings can clearly be seen on the screen.

a To produce an acceptable diffraction pattern the wavelength of the electrons needs to be about 10^{-11} m. Show that the speed of electrons with a de Broglie wavelength of 2.5×10^{-11} m is about 3×10^7 m s^{-1}. **[2]**

b Calculate the size of the accelerating voltage in the electron gun to produce electrons of this speed. **[3]**

c Explain why electrons with this wavelength are needed. **[2]**

9 At the beginning of the twentieth century there were two competing models of the atom – Thomson's plum pudding model and Rutherford's nuclear model.

a Outline the essential features of each model. **[4]**

In 1909 Geiger and Marsden carried out a series of experiments in which they fired α-particles at thin gold foils.

b i Outline the results obtained by Geiger and Marsden. **[3]**

ii Explain how these results provided conclusive evidence that the plum pudding model was incorrect. **[2]**

Unit 4: Practice unit test

Section A

1 Which of the following pairs of quantities *must* be conserved in a particle decay?
A charge and mass
B charge and momentum
C energy and mass
D energy and momentum **[1]**

2 In detecting charged particles, all particle detectors rely upon the process of
A annihilation
B excitation
C ionisation
D scattering **[1]**

3 A football falls to a level surface with a momentum 1.5 N s. After impact it rebounds with a momentum 0.3 N s. The ball's change in momentum is
A 1.8 N s away from the ground
B 1.8 N s towards the ground
C 1.2 N s away from the ground
D 1.2 N s towards the ground **[1]**

4 A capacitor is charged with a 6 V battery. The capacitor is disconnected, then connected to a 12 V battery with the same polarity. The ratio

$$\frac{\text{energy stored charging from 6 V to 12 V}}{\text{energy stored charging from 0 V to 6 V}}$$

A is equal to 1
B is greater than 1
C is less than 1
D is impossible to calculate **[1]**

5 The charge-to-mass ratio of an electron $\left(\frac{e}{m}\right)$ is equal to 1.76×10^{11} C kg^{-1} at low electron speeds. At speeds approaching 3×10^{8} m s^{-1}, $\frac{e}{m}$ is
A greater than 1.76×10^{11} C kg^{-1} because the charge increases
B greater than 1.76×10^{11} C kg^{-1} because the mass decreases
C less than 1.76×10^{11} C kg^{-1} because the charge decreases
D less than 1.76×10^{11} C kg^{-1} because the mass increases **[1]**

6 Which one of the following is *not* composed of quarks?
A muons
B neutrons
C pions
D protons **[1]**

7 In an elastic collision the particles do *not* have to:
A conserve kinetic energy
B conserve momentum
C make a head-on impact
D rebound after the impact **[1]**

8 When a tilting train takes a corner at speed, the train leans in towards the centre of its circular path. The train tilts in order to
A bring the train to equilibrium
B give an extra centripetal force
C obey Newton's third law
D oppose the centrifugal force **[1]**

9 Which of the following *cannot* be units of voltage?
A JC^{-1}
B NmC^{-1}
C Tms^{-1}
D Wbs^{-1} **[1]**

10 Which of the following charges is *not* possible for a meson? See the data opposite.
A −1
B 0
C +1
D +2 **[1]**

	Quarks		Leptons	
	u	d	e^-	ν_e
	c	s	μ^-	ν_μ
	t	b	τ^-	ν_τ
$\frac{\text{Charge}}{e}$	$\frac{2}{3}$	$-\frac{1}{3}$	−1	0

[10 marks]

Section B

Quality of written communication will be tested in questions marked with an asterisk.

***11** A teacher demonstrates Faraday's law by bringing a magnet near to a solenoid connected to a centre-zero micro-ammeter. When the magnet is gently inserted into one end of the solenoid there is a small current recorded. When the magnet is inserted at higher speed the current is larger. Removing the magnet from the solenoid produces a current in the opposite direction.

Explain these observations, making reference to Faraday's and Lenz's laws. **[5]**

12 a A 470 μF capacitor is charged to a pd of 6.0 V. Calculate:
i the charge stored on the capacitor, **[2]**
ii the energy stored by the capacitor. **[2]**
b Describe how you would show experimentally that the pd across a 470 μF capacitor is proportional to the charge on the plates of the capacitor. **[3]**

***13** At the start of the twentieth century Rutherford initiated a series of experiments that proved pivotal to establishing the model of the atom. When Geiger and Marsden reported α-particles rebounding from the gold foil, Rutherford was astounded. 'It was quite the most incredible event that ever happened to me in my life. It was almost as incredible as if you fired a 15-inch shell at a piece of tissue paper and it came back and hit you.' Why was Rutherford surprised at the scattering results, and how did he explain them? **[5]**

14 A student carries out an experiment to investigate momentum and collisions. A ball bearing is rolled down a ramp clamped at the edge of a bench. The ball bearing makes a head-on collision with a marble at the bottom of the ramp.

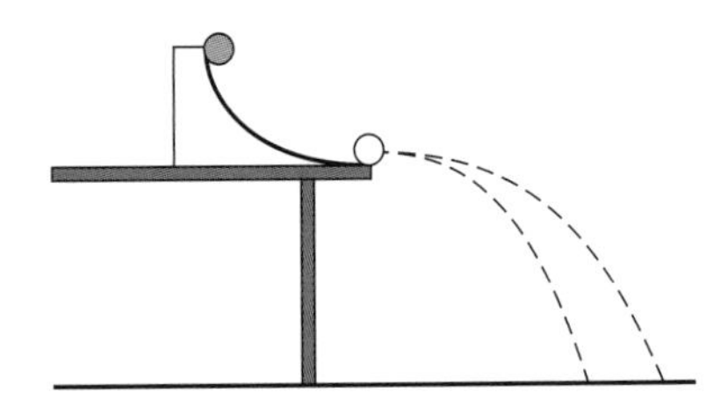

The ball bearing and the marble both move forward in the initial direction of the ball bearing after the collision, landing in a tray containing a layer of sand.
a Initially the ball bearing rolls down the ramp without the marble in place. The ball bearing falls a vertical distance of 1.2 m and lands a horizontal distance 95 cm forward of the end of the ramp. Show that the speed of the ball bearing leaving the ramp is about $2\,ms^{-1}$. **[4]**

b When the ball bearing is rolled down the ramp with the marble in place, it knocks the marble forwards. The ball bearing lands a horizontal distance 64 cm forward of the end of the ramp, and the marble lands a further 93 cm forward from this position. Calculate the ratio of masses of the ball bearing and the marble. **[3]**

c The student suspects that the collision between the ball bearing and the marble is elastic. Explain what is meant by an 'elastic' collision and determine whether the student's hypothesis is correct. **[3]**

15 In Formula One racing, cars often take corners at speeds in excess of 50 m s^{-1}, and so there is a very real danger of skidding.

a i In older race tracks the corners are banked. Explain why banking of the track enables cars to take the corner at greater speeds. **[1]**

ii At a corner of radius 260 m the track is banked at 20°. Calculate the maximum speed if no frictional force is required. **[4]**

b 'Downforce' is used to keep a car in contact with the track. The motion through the air produces a force perpendicular to the direction of travel, pushing the car onto the track. A racing car of mass 720 kg takes an unbanked corner of radius 550 m at a speed of 50 m s^{-1}. The maximum frictional force is 0.4 × (reaction from track). Calculate the downforce necessary to prevent the car from skidding outwards. **[4]**

16 A teacher builds a simulation of a particle detector by sprinkling fine sand into the lid of a shoebox. She rolls a marble down a ramp placed at one end.

a Describe the path of the marble from when it leaves the ramp. **[1]**

The teacher attaches a magnet to the underside of the lid, and a steel ball bearing is rolled down the ramp slightly to one side of the magnet, as shown.

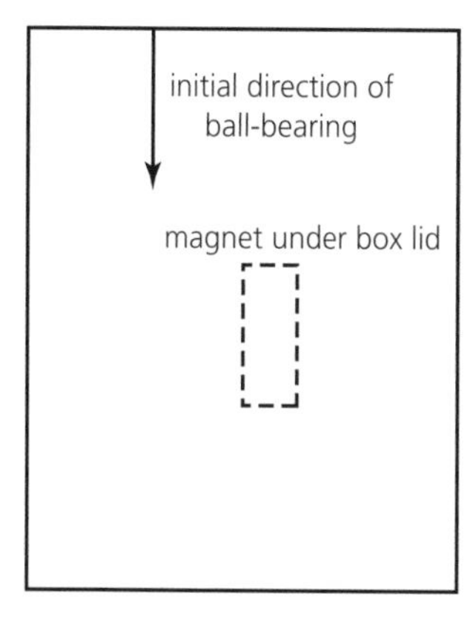

b i Sketch and explain the ball bearing's path in the lid. **[2]**

ii The steel ball bearing is released from higher up the ramp. Sketch and explain the new path. **[2]**

c To what extent do you think that the simulation is a useful analogue to the detection of particles in bubble chamber experiments? **[3]**

17 a A van de Graaff generator is used to produce a large potential difference. The dome of the generator is spherical with a radius of 15 cm. After the generator has been running for a few minutes the dome has accumulated negative charge and the pd between the dome and the Earth is 75 kV.

i If the dome capacitance is 67 pF, calculate the charge stored. **[2]**

ii Calculate the electric field strength close to the dome's surface. **[2]**

iii Sketch the electric field in the region around the dome. **[1]**

b A cloud 1.25 km above the Earth's surface has an approximately rectangular shape. In an electrical storm, charge builds up on the cloud until a uniform electric field of 3.00×10^6 V m^{-1} in the space between the cloud and the Earth makes the air break down and conduct electricity as a lightning bolt.

i Regarding the Earth's surface and the cloud as being the 'plates' of a capacitor of 12 nF, show that the maximum charge stored in the Earth–cloud capacitor is about 50 C. **[3]**

ii Explain why the electric field between the cloud and the Earth can be taken to be uniform, although the Earth's surface is curved. **[1]**

***18** A mass spectrometer is an instrument used to determine the relative masses of ions and hence can be used for isotopic analysis. The atoms in the sample are ionised and then the ions are accelerated and collimated into a fine beam. This ion beam is passed through a magnetic field and the ions are deflected by different amounts according to their mass.

Thomson investigated a sample of neon using an early version of a mass spectrometer, determining that the sample contained two different mass neon atoms. He called these *isotopes* of neon.

a Explain what is meant by the term 'isotope'. **[1]**

b In the mass spectrometer, ions are accelerated through a pd of 10.0 kV to produce a fast ion beam. Show that the speed of singly ionised neon-20 atoms is about $3 \times 10^5 \, m \, s^{-1}$. The mass of an ion of neon-20 is 19.992 u. **[3]**

c In one type of mass spectrometer the ions are passed through a velocity selector. This is a pair of parallel plates 2.5 cm apart. A pd of 140 V is applied to create an electric field. A magnetic field is also applied, so as to give the ions a force in the opposite direction to that from the electric field.

i Calculate the magnetic field strength needed to allow neon-20 ions from part **b** to pass undeflected through the velocity selector. **[3]**

ii Draw a diagram of the velocity selector showing the direction of the electric and magnetic fields. **[2]**

d The ion beam enters a region of uniform magnetic field of strength 0.3 T with the ions travelling at right angles to the field direction. The ions travel in a circular path.

i Explain why the ions travel in a circular path. **[2]**

ii Calculate the radius of this path. **[2]**

iii Sketch the path of the ions in the field. Add another line to show the path of **neon-22 ions** entering the magnetic field with the same velocity. Explain any differences in the paths. **[2]**

e In 1897 Thomson used magnetic and electric field deflection to determine the charge-to-mass ratio of the electron. Less than 30 years later his son demonstrated the diffraction of electrons as they passed through thin metal foils. What do these two experiments tell us about the electron, and what do they tell us about the nature of science? **[3]**

[70 marks]

[Total 80 marks]

Internal energy

All matter consists of particles, often in the form of molecules. The three common states of matter are solid, liquid, and gas. Whichever state the matter is in, its molecules are in random motion. In solids the motion is vibration; in liquids and gases the motion is from one place to another.

The molecules have **kinetic energy** because of their motion. They also have **potential energy**. There are forces of attraction between them, so when they move apart the potential energy between them increases. The sum of the kinetic and potential energies within a sample of matter is known as its **internal energy** (symbol U).

Temperature

Each of the following four points is helpful in understanding temperature.

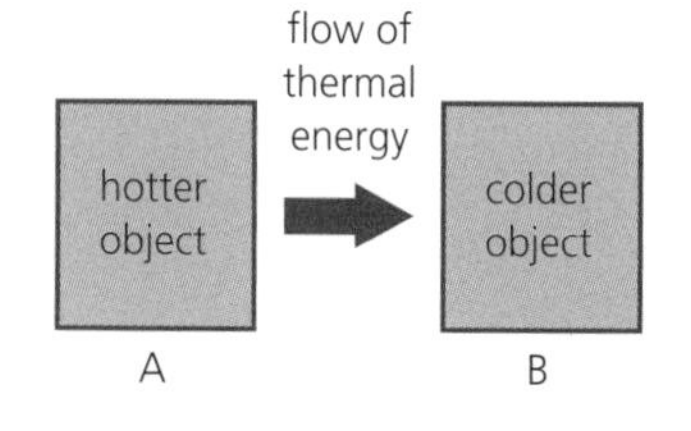

1. If two objects A and B are placed together, and energy (often called **thermal energy** or **heat**) moves from A to B, then A started out at a higher temperature (hotter) than B.
2. Adding thermal energy to an object raises its temperature (so long as it doesn't change state).
3. The molecules in a sample of matter move randomly, with a variety of speeds (see diagram). The temperature of the sample is a measure of the average kinetic energy of its molecules.
4. If you could take all the thermal energy out of an object, its temperature couldn't fall any further. This point leads to the idea that there is an **absolute zero** of temperature.

The **Celsius scale** of temperature is in widespread use, both by scientists and non-scientists. This is defined so that water freezes at 0 °C, and boils at 100 °C.

The absolute zero of temperature is at –273 °C (strictly –273.15 °C, but three significant figures is enough usually). The **absolute** or **Kelvin scale** of temperature has absolute zero as 'zero degrees kelvin', or 0 K. Other Kelvin temperatures are obtained by adding 273 to the Celsius temperature:

$$T/\text{K} = t/°\text{C} + 273$$

Number of particles with speed c
Low temperature
High temperature
Speed c

Distribution of molecule speeds for two different temperatures

This means that an interval of one degree is the same on both Kelvin and Celsius scales.

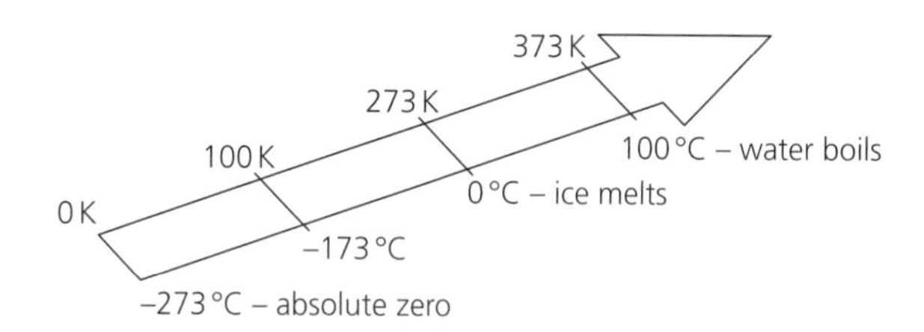

Celsius and Kelvin temperatures compared

Specific heat capacity

The **specific heat capacity** of a material is the energy needed to change the temperature of 1 kg of the material by 1 °C (or 1 K).

$$\Delta E = mc\,\Delta\theta$$

where ΔE is the energy supplied or removed, m is the mass of the sample, c is the specific heat capacity, and $\Delta\theta$ is the temperature change. The unit of c is $\mathrm{J\,kg^{-1}\,°C^{-1}}$ or $\mathrm{J\,kg^{-1}\,K^{-1}}$.

- To find a value for c for a material, we need to devise an experiment in which we can measure ΔE, m, and $\Delta\theta$.
- Using an electrical heater makes it easy to measure ΔE, or the power input (the rate at which energy is supplied).
- To obtain $\Delta\theta$, we might use a temperature sensor and data logger to observe how the temperature rises with time (see graph).
- In a school experiment we might get a value for c accurate to within 10%.
- To get a more accurate value we have to take care to insulate the material. Otherwise it will lose heat to the surroundings during the experiment.
- The accuracy would also be affected because some energy goes into the heater and connections.

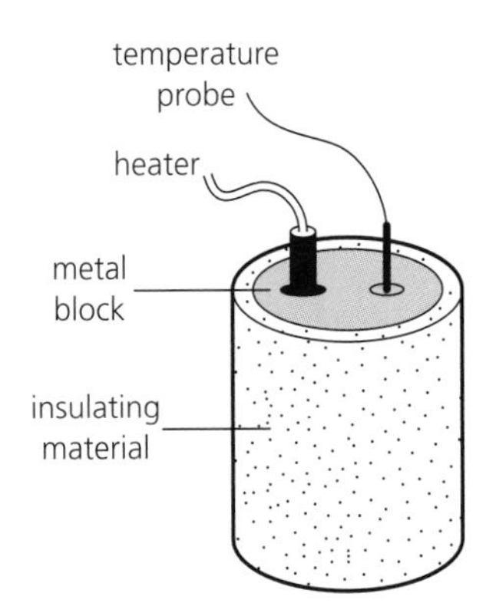

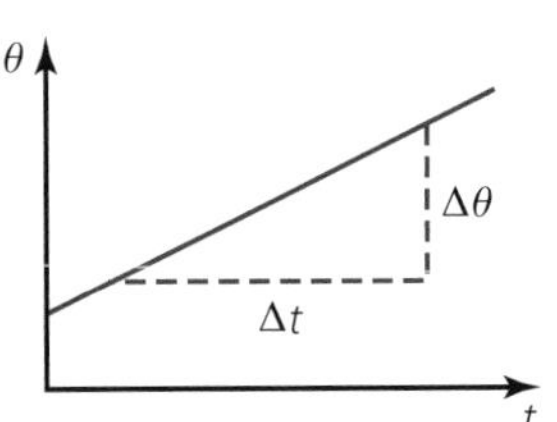

Experiment to measure the specific heat capacity of a metal

Worked Example

In the experiment shown, the heater receives electrical power at 50 W. The mass of the metal block is 0.96 kg. In 240 s (Δt) the temperature rises 14 °C ($\Delta\theta$).

a Calculate a value for the specific heat capacity for the metal.

b Discuss whether your value is likely to be too high or too low.

---

a State the equation you will use; then re-arrange to make the quantity being asked into the subject of the equation.

$$\Delta E = mc\,\Delta\theta \qquad \therefore\ c = \frac{\Delta E}{m\,\Delta\theta}$$

Substitute in the numbers, using the relationship ΔE = power × Δt.

$$c = \frac{50\,\mathrm{W} \times 240\,\mathrm{s}}{0.96\,\mathrm{kg} \times 14\,\mathrm{K}}$$

Now use your calculator to work out the answer, and include the unit.

$$c = 893\,\mathrm{J\,kg^{-1}\,K^{-1}}$$

b Energy may have been lost from the metal block during the process. Also, some energy will have been needed to heat up the heater material and connections. Both these losses would cause $\Delta\theta$ to be too low, which would mean our value for c is too high.

ResultsPlus
Watch out!

Make the quantity you want (in this case specific heat capacity *c*) the subject of the equation *before* you put in the numbers. Also, it's a good habit to put the units in the calculation with the numbers, to remind you in case there's any unit converting to do.

Quick Questions

Q1 **a** Copper melts at a temperature of 1085 °C. What is this temperature in kelvin?
b The supercooled electromagnets at CERN operate at a temperature of 2 K. What is this temperature in degrees Celsius?

Q2 If no energy moves between objects A and B, what can you say about A and B?

Thinking Task

a The specific heat capacity of water ($4200\,\mathrm{J\,kg^{-1}\,K^{-1}}$) is much higher than for almost any other liquid. Why does this make water a good liquid to use in central-heating radiators?

b Estimate how long a 3 kW electric kettle should take to boil the water for a cup of tea.

Gas laws and kinetic theory

The behaviour of gases can be observed by *experiments* in a school lab. This observed behaviour agrees with what is predicted by a *model* called the **kinetic theory**.

Experimental gas laws

The following quantities can be measured directly in experiments:

p – the pressure within the gas (unit Pa, or N m^{-2})
V – the volume of space occupied by the gas (unit m^3)
T – the kelvin temperature of the gas (unit K).

Each experiment is designed to keep one of these three quantities fixed (constant), and to measure how the other two quantities are related. The results of these experiments are:

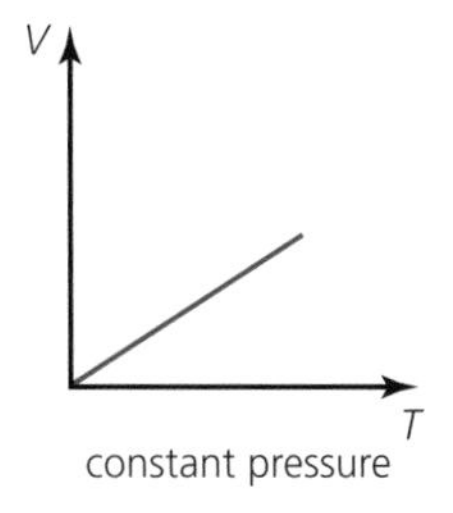

constant pressure

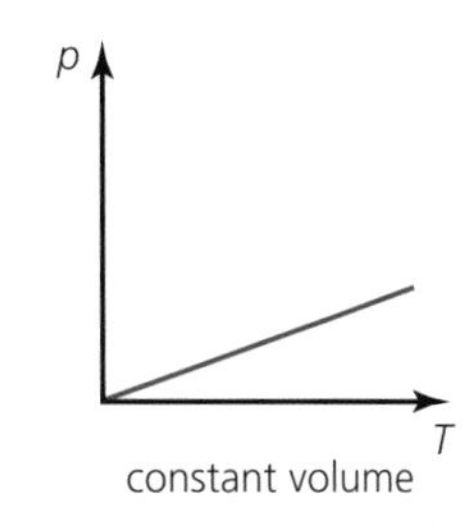

constant volume

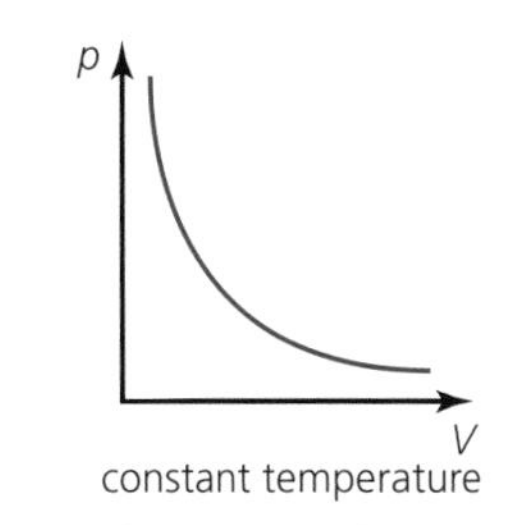

constant temperature

Temperature, volume and pressure relationships for a fixed sample (fixed mass) of gas

1. With p constant, $V \propto T$
2. With V constant, $p \propto T$
3. With T constant, $p \times V$ = constant, or $V \propto 1/p$, or $p \propto 1/V$

Gas equations

These results can be summarised in two equations. First,

$$pV = NkT$$

where N is the number of particles in the mass of gas, and k is the Boltzmann constant (k = 1.38 × 10^{-23} J K^{-1}). This equation is sometimes called the **ideal gas equation**. An ideal gas would be one that perfectly obeys this equation in all circumstances. In reality most gases obey it under most conditions, and in exam questions you can assume that it can be used.

Thinking Task

The second equation follows from the first. Can you show that? Hint: a fixed mass of gas means N is constant.

The second equation combines all three experimental laws into one equation:

$$\frac{p_2 V_2}{T_2} = \frac{p_1 V_1}{T_1}$$

This refers to a fixed mass of gas when two or even three of p, V and T change from values 1 (subscript 1) to values 2 (subscript 2).

ResultsPlus Watch out!

You must convert Celsius temperatures to kelvin when using these equations. Thousands of candidates every year lose marks by forgetting to do this. Sometimes examiners give a clue because the numbers in Celsius are rather odd, e.g. 27 °C. But don't rely on this prompt!

Worked Example

A closed container of gas is initially at atmospheric pressure (1.00 × 10^5 Pa) and a temperature of 27 °C. What is the pressure in the container when it is heated to 267 °C?

The volume is fixed so $V_1 = V_2$. So the equation becomes

$$\frac{p_2}{T_2} = \frac{p_1}{T_1} \therefore p_2 = \frac{p_1 T_2}{T_1}$$

Substitute and calculate:

$$p_2 = \frac{1.00 \times 10^5\,\text{Pa} \times (267 + 273)\,\text{K}}{(27 + 273)\,\text{K}} = \frac{1.00 \times 10^5\,\text{Pa} \times 540\,\text{K}}{300\,\text{K}} = 1.80 \times 10^5\,\text{Pa}$$

Kinetic theory of gases

You can use Newton's laws about forces and momentum to model the behaviour of a gas, but you need to start with some assumptions about the gas molecules:

- the molecules of the gas are in rapid random motion
- the molecules make perfectly elastic collisions with each other and with the container walls
- the molecules exert no forces on each other or the walls except during collisions
- the molecules occupy negligible volume compared with the volume of the container
- the time spent in collisions is negligible compared with the time between collisions.

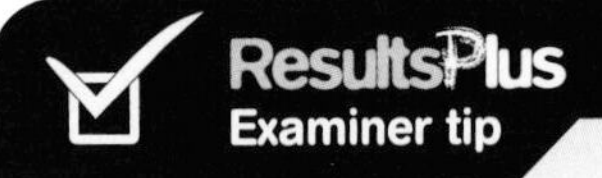

You do not need to learn these assumptions for the exam. However you will not be able to discuss the kinetic theory unless you understand at least the first two points.

These assumptions describe how an ideal gas would behave. From this starting point, the following equation for an ideal gas can be deduced:

$$\frac{1}{2}m\langle c^2\rangle = \frac{3}{2}kT$$

where m is the mass of one molecule, c is the speed of a molecule, and $\langle c^2\rangle$ is the mean (average) value of c^2 for all the molecules. Thus both sides of this equation represent the average kinetic energy of the molecules, and the theory confirms our idea that temperature is a measure of this average kinetic energy. The 'root mean square' (rms) speed is the square root of $\langle c^2\rangle$.

Worked Example

a Five gas molecules have speeds of 300 m s^{-1}, 450 m s^{-1}, 520 m s^{-1}, 680 m s^{-1} and 730 m s^{-1}. Calculate the rms speed for these molecules.

b The molecules of a real gas have both kinetic and potential energy. Explain why an ideal gas has only kinetic energy.

Edexcel June 2008 Unit Test 2

---

a Calculate the square of each speed, in (m s^{-1})^2:

90 000, 202 500, 270 400, 462 400, 532 900

Find the average of the squared values: 311 640 (m s^{-1})^2 and take the square root of this average value: 558 m s^{-1}.

b One of the five assumptions about an ideal gas is that the molecules exert no forces on each other. Therefore there is no potential energy between them, and all the internal energy is kinetic energy.

Quick Questions

Q1 Starting from the ideal gas equation, show that the unit of k is J K^{-1}.

Q2 An ideal gas is heated from 300 K to 600 K. By what factor does the average speed of its molecules change?

Q3 The volume of air in a car tyre is 0.02 m^3 when the pressure in the tyre is 3.5 atmospheres (3.5×10^5 Pa). What volume would this air occupy when released to atmospheric pressure?

In some ways the behaviour of snooker balls on a table resembles that of gas molecules. How well does each of the five assumptions about gas molecules apply to snooker balls? (very well/pretty well/not well at all)

Section 4: Thermal energy checklist

By the end of this section you should be able to:

Revision spread	Checkpoints	Spec. point	Revised	Practice exam questions
Internal energy	Explain the concept of internal energy as the random distribution of potential and kinetic energy among molecules.	110		
	Explain the concept of absolute zero and how the average kinetic energy of molecules is related to the absolute temperature.	111		
	Recognise and use the expression $\Delta E = mc\,\Delta\theta$.	109		
Gas laws and kinetic theory	Recognise and use the expression $\frac{1}{2}m\langle c^2\rangle = \frac{3}{2}kT$.	112		
	Use the expression $pV = NkT$ as the equation of state for an ideal gas.	113		

ResultsPlus
Build Better Answers

In July 2003 there was an attempt to fly a manned, spherical balloon to a height of about 40 kilometres. At this height the atmospheric pressure is only one thousandth of its value at sea level and the balloon would have expanded to a diameter of 210 m. The temperature at this height is –60 °C. The attempt failed because the thin skin of the balloon split while it was being filled with helium at sea level.

Make an estimate of the temperature at sea level, and hence obtain the volume of helium the balloon would have contained at sea level if it had been filled successfully. **[6]**

Edexcel June 2007 Unit PSA5

Examiner tip

A question like this has a lot of steps. There will be 1 mark or more for each step, so even if you can't immediately see where it's going, you need to start by writing something down – in this case using the hint and the data they've given you. That gives you a better chance of spotting which equation you need to solve it.

Student answer	Examiner comments
$\frac{p_2V_2}{T_2} = \frac{p_1V_1}{T_1}$ so $V_1 = \frac{p_2V_2T_1}{T_2p_1}$	This is the right way to start. Show each step in your thinking and calculating. Even if your final answer is a long way off, the marker can see your steps, and give you some marks for sensible ones.

Student answer	Examiner comments
Pressure is p_1 at sea level, and $p_2 = 0.001 \times p_1$ V_2 is a sphere 210 m in diameter; so $V_2 = \frac{4}{3}\pi r^3 = \frac{4}{3}\pi \times \left(\frac{210}{2}\right)^3 = 4.8 \times 10^6\,\text{m}^3$ Estimate sea level temperature as 20 °C, so $T_1 = (20 + 273)\,\text{K} = 293\,\text{K}$ $T_2 = (-60 + 273)\,\text{K} = 213\,\text{K}$	This is good. Several of the quantities in the equation are quite complex, so it's a good idea to write each of them out separately before substituting them into the equation.

Examiner tip

This question contains two common pitfalls: remembering Celsius to kelvin; and spotting diameter not radius.
Be on your guard for both of these!

Student answer	Examiner comments
$V_2 = \frac{p_2V_2T_1}{T_2p_1} = \frac{0.001 \times p_1 \times 4.8 \times 10^6\,\text{m}^3 \times 293\,\text{K}}{213\,\text{K} \times p_1} = 6.6 \times 10^3\,\text{m}^3$	This answer would get full marks.

Practice exam questions

1 A valid set of units for specific heat capacity is
A $kg\,J^{-1}\,K^{-1}$ **B** $kg\,J\,K^{-1}$ **C** $kg^{-1}\,J\,K^{-1}$ **D** $kg\,J^{-1}\,K$ **[1]**

2 A large gas holder, of fixed volume, is made up of sheets of metal riveted together. The pressure of gas in the holder is raised from 1×10^5 Pa to 5×10^5 Pa, while the temperature of the gas remains the same. Which of these statements is true?
A Each sheet is now struck by fewer molecules.
B The mass of gas in the holder is now five times as great.
C The average speed of the molecules has increased by a factor of five.
D The force exerted by the gas on each sheet has not changed. **[1]**

3 The graphs show the distributions of kinetic energy of the molecules in the atmosphere, at sea level and at 40 km above sea level, where it is much colder. Label the sea level graph and give a reason for your answer. **[2]**
Edexcel June 2007 Unit PSA5ii (modified)

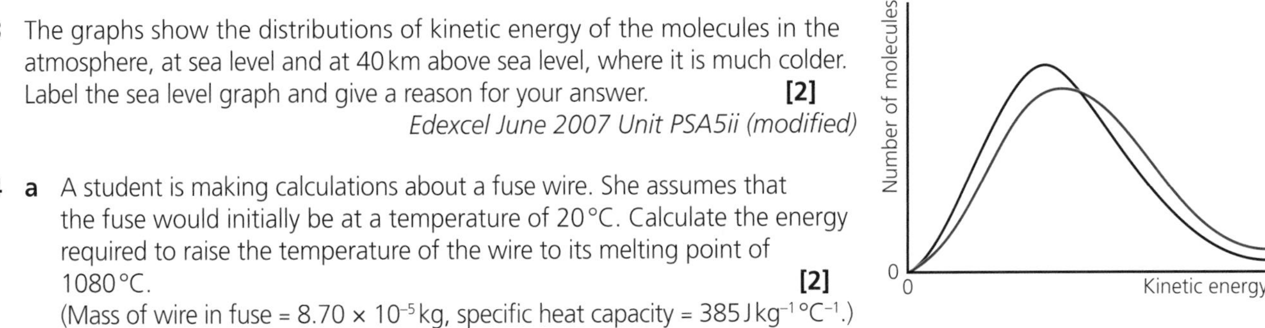

4 a A student is making calculations about a fuse wire. She assumes that the fuse would initially be at a temperature of 20 °C. Calculate the energy required to raise the temperature of the wire to its melting point of 1080 °C. **[2]**
(Mass of wire in fuse = 8.70×10^{-5} kg, specific heat capacity = $385\,J\,kg^{-1}\,°C^{-1}$.)
b Calculate the time for the wire in the fuse to reach its melting point. The electrical power going into the fuse is 2.2 W. **[1]**
c Discuss whether this is likely to be the actual time for the fuse to reach its melting point. **[2]**
Edexcel June 2007 Unit PSA1 (modified)

5 a What is meant by the absolute zero of temperature? **[1]**
b i The Football Association rules require a football to have a maximum volume of $5.8 \times 10^{-3}\,m^3$ and a maximum pressure of 1.1×10^5 Pa above atmospheric pressure (1.0×10^5 Pa). Assuming that the thickness of the material used for the ball is negligible and that the air inside the ball is at a temperature of 10 °C, calculate the maximum number of molecules of air inside the football. **[4]**
ii A football is also required to have a minimum pressure 0.6×10^5 Pa above atmospheric pressure. Assuming the volume of the football remains constant, and that it was initially filled to maximum pressure at 10 °C, calculate the lowest temperature to which the air inside this ball could fall while still meeting the pressure requirements. **[3]**
Edexcel June 2008 Unit Test 2

6 In a radio programme about space tourism, the presenter says that the Earth's atmosphere stops 100 km above the surface. A student decides to put this claim to the test, initially applying the following equation to gas molecules at this height:

$$\frac{1}{2}m\langle c^2\rangle = \frac{3}{2}kT$$

where k is the Boltzmann constant.
a State the meanings of the other symbols used in the equation: m, $\langle c^2\rangle$, T. **[3]**
b What physical quantity does each side of the equation represent? **[1]**
c Calculate a value for the velocity of an oxygen molecule at this height, where the temperature is −50 °C. Mass of oxygen molecule = 5.4×10^{-26} kg. **[2]**
Edexcel June 2008 Unit PSA5ii (modified)

Nuclear radiation

Nuclear radiation consists of particles or gamma rays emitted from the unstable, i.e. radioactive, nucleus of an atom when it decays (see page 34). Each type of radiation emerges with energy, and each **ionises** matter it passes through. This combination of energy and ionisation is what causes damage to living tissue. It also makes it possible to detect the radiation.

Most atoms of most elements are stable, but all elements have some unstable isotopes. These give rise to **background radiation**. This background radiation is not harmful (indeed we have evolved because of the mutations it causes). Human use of radioactive materials has added negligible amounts of radiation to the natural background level. However we need to be extremely careful in the decisions we make and how we handle radioactive materials in the future.

ResultsPlus Examiner tip

Particles also come from the Sun and other cosmic sources, and contribute to the overall background radiation.

α, β and γ radiations

Nuclear radiations are classified as **alpha** (α), **beta** (β) and **gamma** (γ). We can distinguish between them by their differing penetrating powers, by stopping them using absorbing material. A simple experimental arrangement uses a source of radiation, absorbing material, a Geiger-Müller (GM) tube and a counter or data logger.

	α	β	γ
What is it?	helium nucleus $^4_2$He	electron $^0_{-1}$e	photon
What typically stops it?	thin paper, dead skin	a few mm of solid matter, e.g. metal	a few cm of dense metal, e.g. lead
Ionising ability	high	medium	very low
Some uses	smoke detector	carbon dating, paper thickness control	medical or industrial tracers, sterilising surgical equipment

Equations for nuclear decay

Nuclear decay occurs spontaneously, and it is random. Nuclear decay can be modelled by representing an unstable nucleus as a single dice – if it falls with a particular side upward that means it decays. The behaviour of one dice (or nucleus) is unpredictable, but the overall behaviour of a large collection of dice (or nuclei) becomes predictable.

The **activity** A of a sample of radioactive material is the number of decays occurring per second, or the rate of decay. The unit of A is the becquerel (Bq). 1 Bq = 1 disintegration per second.

An important property of a particular isotope is its **half life**, $t_{1/2}$. This it the time it takes for half the unstable nuclei in a sample to decay. It is also the time for the activity to fall to half of its initial value. These equations relate the half life and activity:

$$A = \frac{-\mathrm{d}N}{\mathrm{d}t} = \lambda N$$

$$N = N_0\,\mathrm{e}^{-\lambda t} \text{ and } \lambda = \frac{\ln 2}{t_{1/2}}$$

where N is the number of unstable nuclei left in the sample at a particular moment; N_0 is the initial number of unstable nuclei; and λ is the **decay constant** for the particular isotope.

Minus signs appear in these equations because the number of unstable nuclei is decreasing. In means the same as $\log_{10}$e.

Worked Example

a The isotope $^{131}_{53}$I has a half life of 8 days. One sample has an initial activity of 80 MBq. Sketch a graph showing how the activity changes with time for the next 24 days.

b Another sample of the isotope $^{131}_{53}$I has an initial activity of 56 MBq. Calculate the time for its activity to fall to 20 MBq.

Edexcel June 2001 Unit PSA1

a

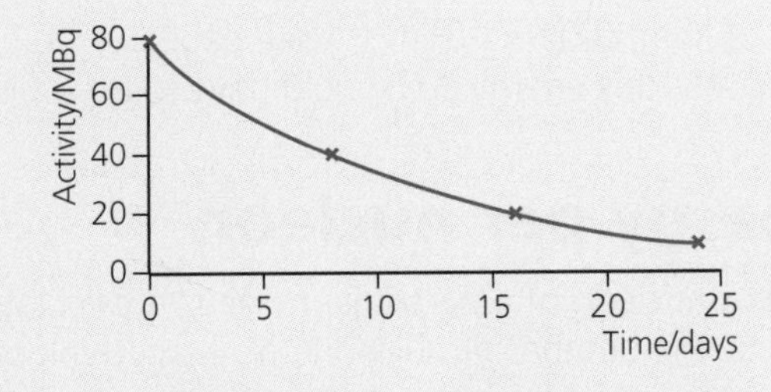

b First calculate λ: $\lambda = \frac{\ln 2}{t_{1/2}} = \frac{\ln 2}{8 \text{ days}} = 0.0866 \text{ days}^{-1}$

Now you need this equation:

$$N = N_0 e^{-\lambda t}$$

Note carefully the maths used to make t the subject:

$$\frac{N}{N_0} = e^{-\lambda t}$$

$$\ln\left(\frac{N}{N_0}\right) = -\lambda t$$

So $$t = \ln\left(\frac{N}{N_0}\right) \div \lambda$$

$$= \ln\left(\frac{20 \text{ MBq}}{56 \text{ MBq}}\right) \div 0.0866 \text{ days}^{-1} = 11.9 \text{ days}$$

Quick Questions

Q1 Which of these has the higher activity: 2×10^{20} nuclei of isotope X, with half life 4×10^{6} years, or 5×10^{9} nuclei of Y, with half life 6 hours?

Q2 The half life of $^{16}_{7}$N is 7.1 s

a Calculate the decay constant for $^{16}_{7}$N.

b Show that the time taken for the mass of $^{16}_{7}$N in a sample to decrease from 5.0 μg to 1.0 μg is approximately 16 s.

ResultsPlus Watch out!

Even though you are asked for a 'sketch', make sure your graph shows as much detail as possible. It must show that the activity halves every 8 days. Plot points every 8 days to show when the activity is 40, 20 and 10 MBq, then join them with a smooth curve.

ResultsPlus Watch out!

There are several points to note about this calculation.

1. Units for t and λ. This example uses days and days^{-1} throughout. Always stick to the same time unit throughout one of these calculations.
2. The equation refers to N and N_0. When substituting numbers we have used $\frac{A}{A_0}$ instead of $\frac{N}{N_0}$. That works, since A is proportional to N.
3. At the end of any calculation, ask yourself whether the answer makes sense. In this case the activity fell to less than half the initial activity, and we have ended with an answer of rather more than the half life – so it does make sense.

Thinking Task

a What is the link between a radiation's ionising ability and its ability to penetrate material?

b In what ways are the behaviours of a collection of dice and a sample of radioactive isotopes similar?

Binding energy, fission and fusion

The mass of any nucleus is slightly less than the total mass of the separate nucleons. This difference is called the **mass defect** Δm.

As separate nucleons become bound together to form a nucleus they lose energy. This lost energy is called the **binding energy** ΔE and is related to the mass defect:

$$\Delta E = c^2 \Delta m$$

where c is the speed of light, $3.00 \times 10^8 \text{ m s}^{-1}$ (see page 36). To pull a nucleus apart into its separate nucleons, this energy ΔE would have to be supplied.

Binding energy per nucleon

The binding energy per nucleon of a particular nucleus is calculated by dividing the total binding energy by the number of nucleons. You can think of this as the energy needed to remove each nucleon from the nucleus if it were pulled apart.

Nuclear masses are often expressed in atomic mass units, u (see page 37). $1 \text{ u} = 1.66 \times 10^{-27} \text{ kg}$.

Proton mass $m_p = 1.00728 \text{ u}$. Neutron mass $m_n = 1.00867 \text{ u}$.

Binding energies are often expressed in electronvolts, eV, or MeV. $1 \text{ eV} = 1.60 \times 10^{-19} \text{ J}$; $1 \text{ MeV} = 1.60 \times 10^{-13} \text{ J}$.

ResultsPlus Watch out!

With these calculations you need to keep all the significant figures until you have done the subtraction in step 1. Also, beware of confusing eV and MeV.

Worked Example

The nucleus $^{10}_{5}\text{B}$ has mass 10.01294 u. Calculate its mass defect and binding energy. Express the binding energy in MeV per nucleon.

$^{10}_{5}\text{B}$ has proton number 5, so it contains 5 protons. It has 10 nucleons in total, so it must also have 5 neutrons.

mass of separate nucleons $= 5 \times m_p + 5 \times m_n$

$= 5 \times 1.00728 \text{ u} + 5 \times 1.00867 \text{ u} = 10.07975 \text{ u}$

mass defect $\Delta m = 10.07975 \text{ u} - 10.01294 \text{ u} = 0.06681 \text{ u}$

$= 0.06681 \times 1.66 \times 10^{-27} \text{ kg} = 1.11 \times 10^{-28} \text{ kg}$

Calculate the binding energy in joules:

$$\Delta E = c^2 \Delta m = (3.00 \times 10^8 \text{ m s}^{-1})^2 \times 1.11 \times 10^{-28} \text{ kg} = 9.98 \times 10^{-11} \text{ J}$$

Convert this to MeV:

$$9.98 \times 10^{-11} \text{ J} = \frac{9.98 \times 10^{-11} \text{ J}}{1.60 \times 10^{-13} \text{ J MeV}^{-1}} = 62.4 \text{ MeV}$$

$$\therefore \text{ binding energy per nucleon for } ^{10}_{5}\text{B} = \frac{62.4 \text{ MeV}}{10} = 6.24 \text{ MeV}$$

The diagram shows how the binding energy per nucleon depends on the size of the nucleus. The nucleus of $^{56}_{26}\text{Fe}$ has the highest binding energy per nucleon. Its nucleons are the most tightly bound together.

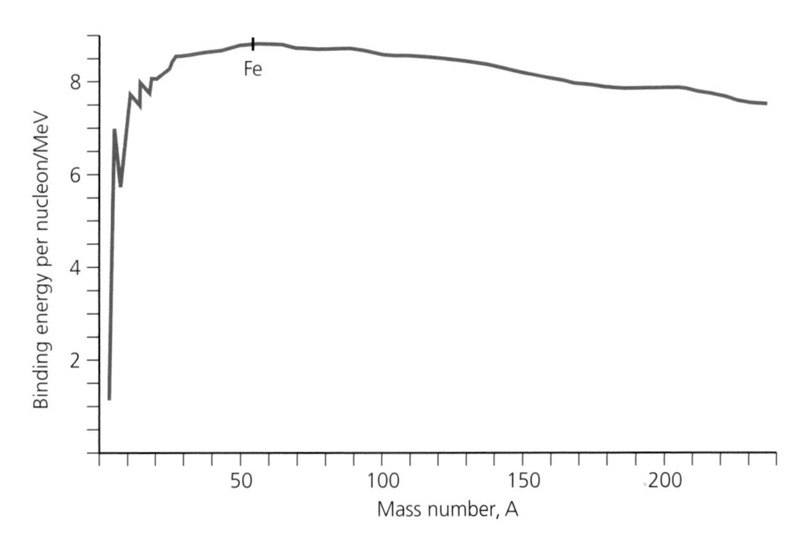

Graph of binding energy per nucleon against mass number

Fusion

When two light nuclei undergo **fusion** they join to make a single nucleus. The average binding energy per nucleon increases and the nucleus loses mass.

As an example, the first stage in the fusion process inside a star is when two protons combine:

$${}^{1}_{1}H + {}^{1}_{1}H \rightarrow {}^{2}_{1}D + {}^{0}_{1}e + v_e$$

As nuclei have positive charge, a lot of energy is needed in the first place to push them together against the force of electrostatic repulsion. For nuclear fusion to take place, temperatures of about 10^7 K or higher are needed to provide the nuclei with enough kinetic energy, and high densities are required so that they collide frequently. These conditions exist in the cores of stars.

Fission

Nuclear **fission** involves a heavy nucleus splitting to form two lighter nuclei (and usually some neutrons). The average binding energy per nucleon increases so there is an overall loss of mass.

In both fusion and fission the mass loss re-appears as energy, in the kinetic energy of the resulting particles and as photons. In nuclear power stations fission is controlled, and the energy is used to generate electricity. In a bomb the fission and resulting uncontrolled chain reaction are extremely rapid.

Quick Questions

Q1 The nucleus ${}^{54}_{26}Fe$ has mass 53.93962 u. Calculate its binding energy in MeV per nucleon.

Q2 For each of the following statements, say whether it applies to **A** nuclear fusion, **B** nuclear fission, **C** both, or **D** neither.

- **a** Two light nuclei join together.
- **b** One heavy nucleus splits apart.
- **c** The event results in an overall loss of mass.
- **d** The event leads to an increase in binding energy per nucleon.
- **e** The event leads to an overall reduction in binding energy.
- **f** The event causes a release of kinetic energy and/or emission of radiation.
- **g** For this to occur requires high temperature and density.

ResultsPlus Examiner tip

When you write about fission and/or fusion, make sure you spell these words accurately. If you don't, there is ambiguity which may lose you marks.

Thinking Task

The main nuclear fusion reactions at the Sun's core are summarised by this equation:

$$4\,{}^{1}_{x}H \rightarrow {}^{y}_{2}He + 2\,{}^{0}_{z}e^{+} + 2\nu_e$$

where e^+ is a positive electron (a positron).

- **a** Fill in the missing numbers x, y, z.
- **b** Calculate the energy released by this fusion of 1 kg of hydrogen nuclei ($6.02 \times 10^{26+}$ nuclei). Mass of He^+ nucleus = 4.00260 u. Mass of e = 0.00055 u. Treat the ν_e as having zero mass.

Section 5: Nuclear decay checklist

By the end of this section you should be able to:

Revision spread	Checkpoints	Spec. point	Revised	Practice exam questions
Nuclear radiation	Show an awareness of the existence and origin of background radiation, past and present.	114		
	Investigate and recognise nuclear radiations (alpha, beta and gamma) from their penetrating power and ionising ability.	115		
	Describe the spontaneous and random nature of nuclear decay.	116		
	Determine the half-lives of radioactive isotopes graphically.	117		
	Recognise and use the expressions for radioactive decay: $\frac{dN}{dt} = -\lambda N$ $\lambda = \ln 2/t_{1/2}$ $N = N_0 e^{-\lambda t}$	117		
	Discuss the applications of radioactive materials, including ethical and environmental issues.	118		
Binding energy, fission and fusion	Explain the concept of nuclear binding energy.	136		
	Recognise and use the expression $\Delta E = c^2 \Delta m$	136		
	Use the non-SI atomic mass unit (u) in calculations of nuclear mass (including mass deficit) and energy.	136		
	Describe the processes of nuclear fusion and fission.	137		
	Explain the mechanism of nuclear fusion and the need for high densities of matter and high temperatures to bring it about and maintain it.	138		

ResultsPlus
Build Better Answers

1 Radioactivity involves the *spontaneous* emission of *radiation* from *unstable* nuclei. Explain the meaning of the words in italics as they apply to the process of radioactivity. **[3]**

Edexcel June 2007 Unit Test 1

Examiner tip

It's really important to use precise words in answers to these sorts of questions. The marker will be looking for specific words for awarding a mark. Write more rather than less – that way you are more likely to include words on the examiner's key words list.

Student answer	Examiner comments
spontaneous: Unaffected by external circumstances such as temperature.	This is correct. Be careful – spontaneous does *not* mean the same as random or unpredictable.
radiation: Alpha, beta, gamma.	The mark would be given for naming any one of these types of radiation, or beta$^+$, beta$^-$, or positron.
unstable: The nucleus is likely to disintegrate.	This will get the mark. It is important to refer to the *nucleus*, not to the atom. It is not correct just to say it has 'high energy', nor 'too many particles'.

ResultsPlus
Build Better Answers

2 The graph shows the predicted change in the activity A over a period of just over 300 years of $^{241}_{95}Am$, an isotope of americium used in smoke alarms. One gigasecond, Gs, is equal to 10^9 s. Use the graph to determine the half life of this isotope in years. (Do *not* attempt to extrapolate the graph.) **[3]**

Edexcel June 2007 Unit Test 6

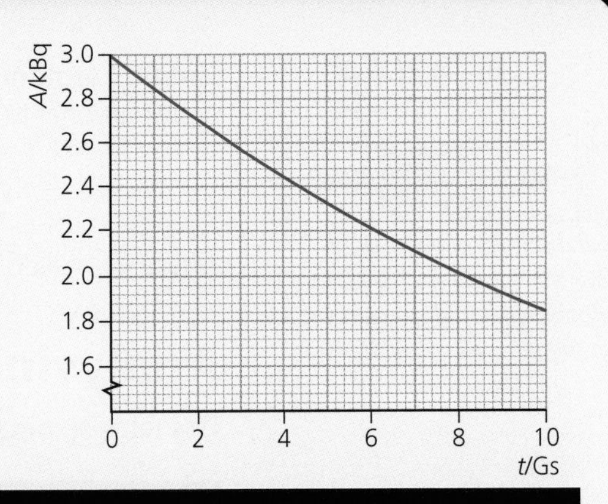

Student answer	Examiner comments
$N = N_0 e^{-\lambda t}$ and $\lambda = \frac{\ln 2}{t_{1/2}}$ $\lambda = \ln\left(\frac{N}{N_0}\right) \div t$ From two points on the graph, at 0 and 8 Gs: $\lambda = \ln\left(\frac{3.0\,\text{kBq}}{2.01\,\text{kBq}}\right) \div 8\,\text{Gs} = 0.0501\,\text{Gs}^{-1}$ $\lambda = \frac{\ln 2}{t_{1/2}}$ so $t_{1/2} = \frac{\ln 2}{\lambda} = \frac{\ln 2}{0.0504\,\text{Gs}^{-1}} = 13.8\,\text{Gs}$	This answer would get 2 out of 3 marks, because the question asks for the half life in years. This question states that it wants the answer in a particular unit – so you must make this conversion for full marks. If the question doesn't specify the unit, then don't attempt any conversion, e.g. a gigasecond answer into seconds, minutes, hours, days, or years. You'll waste valuable time and you may get it wrong.

Practice exam questions

In Questions 1 and 2, choose your answers from the following:

A $^{235}_{92}U + ^{1}_{0}n \rightarrow ^{95}_{38}Se + ^{138}_{54}Xe + 3^{1}_{0}n$

B $^{32}_{15}P \rightarrow ^{32}_{15}S + ^{0}_{-1}e$

C $^{216}_{84}Po \rightarrow ^{212}_{82}Pb + ^{4}_{2}He$

D $^{7}_{3}Li + ^{1}_{1}H \rightarrow ^{4}_{2}He + ^{4}_{2}He$

1 Which of these equations describes the spontaneous decay of a nucleus with the emission of an alpha particle? **[1]**

2 Which of these equations represents a reaction which is used in power stations to provide energy? **[1]**

3 Smoke detectors contain an alpha-emitting source.
a Describe how you would determine whether this radioactive source emits alpha particles only. **[4]**
b State why smoke detectors do not provide a radiation risk in normal use. **[1]**

Edexcel sample assessment material 2007 Unit Test 5

4 Recently, some old human skulls have been found in Mexico. Their age has been established using radiocarbon, ^{14}C, dating.
a When a ^{14}C nucleus decays, it emits a β-particle. State how the composition of the nucleus changes as a result of the decay. **[1]**
b When examining a small sample of one of these old skulls, scientists found that 2.3×10^{-11}% of the carbon was ^{14}C, whereas in recent skulls this proportion is 1.0×10^{-10}%.
i Calculate the age of this old skull. Half-life of ^{14}C = 5730 years. **[3]**
ii Give one reason why the value you calculated above may be inaccurate. **[1]**
iii Recent bones are dated using the decay of ^{210}Pb, which has a half life of 21 years. Explain why ^{210}Pb is more suitable than ^{14}C for dating recent bones. **[1]**

Edexcel June 2008 Unit PSA5

Simple harmonic motion (SHM)

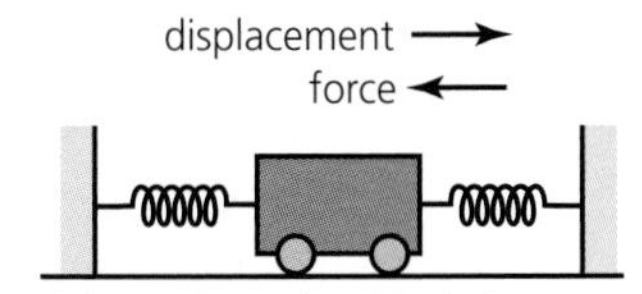

If the springs obey Hooke's law, they give a restoring force proportional to displacement, and so the mass moves with SHM.

Simple harmonic motion (SHM) occurs when the resultant force F acting on an object is proportional to its **displacement** x from equilibrium and in the opposite direction:

$$F \propto -x \quad \text{or} \quad F = -kx$$

where k is called the **force constant** of the system.

Analysing SHM

This table summarises the quantities and symbols involved in SHM.

Quantity	Unit	Description
t	seconds, s	the time which has passed since the motion began being measured (or since $t = 0$)
T	seconds, s	the time for one complete oscillation (constant for a particular SHM situation)
f	hertz, Hz	frequency of the oscillations – number per second (constant for a particular SHM situation)
ω	radians per second, rad s^{-1}	angular frequency of the oscillations (constant for a particular SHM situation)
A	metres, m	amplitude of the oscillations – maximum displacement (constant for a particular SHM situation)
x	metres, m	displacement of the oscillating object from its middle (mean) position (changes as time progresses – therefore a variable)
v	metres/second, m s^{-1}	velocity of the oscillating object (variable)
a	m s^{-2}	acceleration of the oscillating object (variable)
k	N m^{-1}	force constant for system obeying Hooke's law
m	kg	mass of oscillating object

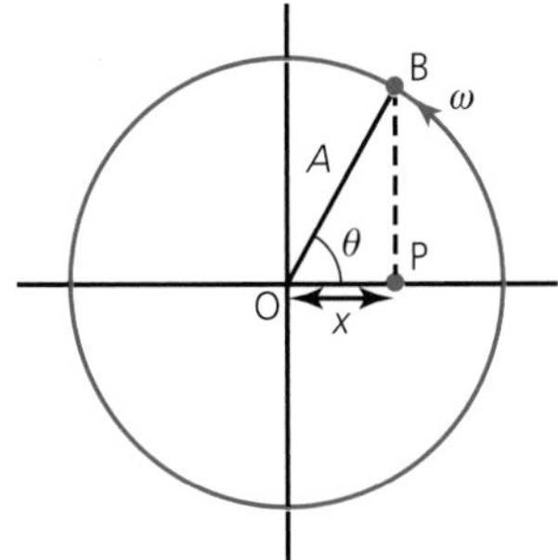

Point P moves with SHM as B moves around the circle. SHM equations are similar to equations for circular motion.

The displacement–time graph for an oscillating object can be obtained using a suitable motion sensor and a data logger. The cosine displacement–time graph typically obtained is shown below.

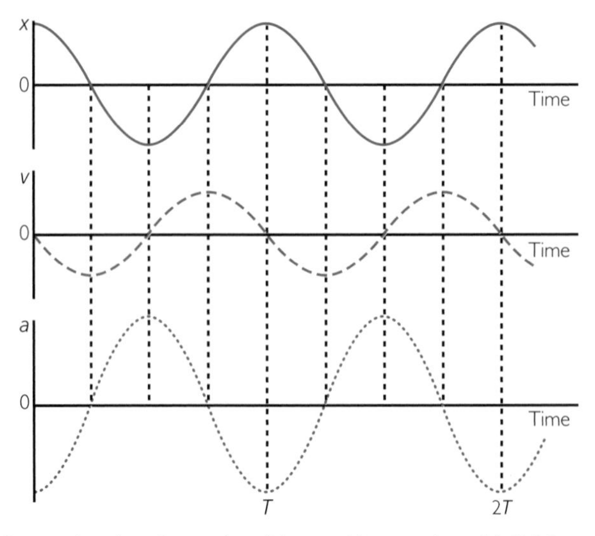

Graphs of x against t, v against t, and a against t for an object moving with SHM

The velocity v at any point can be found from the gradient of the displacement–time graph at that point. The acceleration a can be found from the gradient of the velocity–time graph.

The equations for SHM that you need to recognise and use are given below. These assume that the displacement is maximum at $t = 0$.

$$\omega = \frac{2\pi}{T} \qquad T = \frac{1}{f} = \frac{2\pi}{\omega} \qquad \omega = 2\pi f \qquad x = A \cos \omega t$$

$$v = -A\omega \sin \omega t \qquad a = -A\omega^2 \cos \omega t \qquad a = -\omega^2 x$$

ResultsPlus
Examiner tip

Remembering the following may help you in exams:

- the maximum possible value of $\sin \omega t$ or $\cos \omega t$ is 1
- the maximum velocity is $A\omega$
- the maximum acceleration is $A\omega^2$
- the force constant $k = m\omega^2$.

Worked Example

A mass held between two springs is oscillating with an amplitude of 0.10 m and a period of 2.0 s.

a Calculate the angular frequency ω.
b Calculate the speed of the mass as it passes through the central point of the oscillations.
c The motion begins with the mass at maximum displacement when $t = 0$. Calculate the displacement 2.7 s later.

a $\omega = \frac{2\pi}{T} = \frac{2\pi}{2.0\,\text{s}}\ \text{rad s}^{-1} = 3.14\ \text{rad s}^{-1}$

b At the central point the speed is at its maximum:

maximum velocity $= A \times \omega = 0.10\,\text{m} \times 3.14\,\text{rad s}^{-1} = 0.314\,\text{m s}^{-1}$

c $x = A \cos \omega t = 0.10\,\text{m} \times \cos(3.14\,\text{rad s}^{-1} \times 2.7\,\text{s}) = -0.059\,\text{m}$

ResultsPlus
Watch out!

1. When doing calculations involving ω, $\sin \omega t$, $\cos \omega t$, etc., remember to switch your calculator into radian mode.
2. Take care with signs. The usual convention is that displacements, velocities and accelerations to the *right* are *positive*; a negative value means leftwards.

Quick Questions

Q1 The amplitude of an oscillator is 0.20 m and its period is 3.0 s. Calculate:
a the angular frequency ω,
b the maximum speed,
c the maximum acceleration.

Q2 An oscillator has angular frequency 7.4 rad s^{-1} and amplitude 2.5×10^{-3} m. It has its maximum displacement when $t = 0$. Calculate its displacement, velocity and acceleration when $t = 2.0$ s.

Thinking Task

Explain whether each of these is or is not an example of SHM.

a a person jumping up and down on the ground
b a baby bouncing gently in an elastic harness
c a person trampolining.

Energy and damping in SHM

Energy in SHM

The **kinetic energy** E_k of a simple harmonic oscillator is continuously changing.

- E_k is maximum when speed is maximum (when the displacement $x = 0$).
- E_k is zero when displacement is maximum (the oscillator is momentarily at rest).

$$E_k = \frac{1}{2}mv^2 = \frac{1}{2}m(-A\omega \sin \omega t)^2 = \frac{1}{2}mA^2\omega^2 \sin^2 \omega t$$

The **potential energy** E_p is also changing continuously.

- E_p is maximum when the displacement is greatest.
- E_p is zero as the oscillator moves through its equilibrium position.

$$E_p = \frac{1}{2}kx^2 = \frac{1}{2}kA^2 \cos^2 \omega t = \frac{1}{2}mA^2\omega^2 \cos^2 \omega t$$

If no energy is lost from the oscillator, the sum of its kinetic and potential energies remains constant throughout its motion. This is the **total energy** E_{tot}:

$$E_{tot} = E_k + E_p = \frac{1}{2}kA^2$$

Thinking Task

Explain what happens to the total energy of a mass–spring oscillator if, with everything else remaining the same,

a its amplitude is doubled,
b the mass is doubled,
c the force constant k is doubled (stiffer springs).

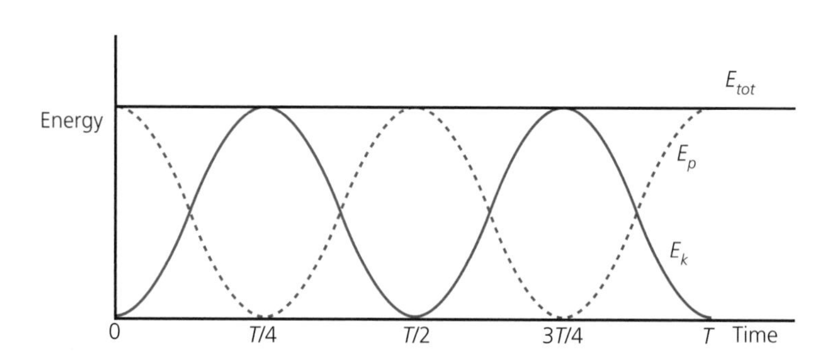

Variation with time of E_k, E_p and E_{tot} for a simple harmonic oscillator

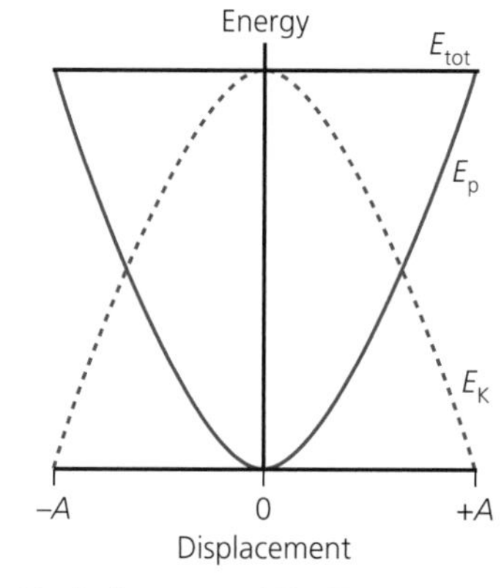

Variation with displacement of E_k, E_p and E_{tot} for a simple harmonic oscillator

Worked Example

A student investigated the behaviour of an oscillating 100 g mass on a spring using a data logger to produce the graph shown.

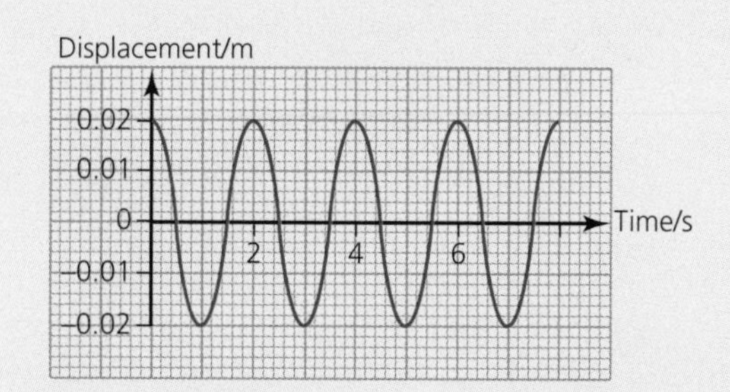

a Write down values for **i** A, **ii** T.

b Calculate the maximum kinetic energy of the 100 g mass.

Edexcel June 2000 Unit PSA6i

If you are asked to write something down, then almost certainly the answer is directly given in the data. But still be careful – it's easy in your excitement to write down the wrong number.

a From the graph, **i** $A = 0.02$ m, **ii** $T = 2.0$ s

b First use the expressions for maximum speed and for ω:

maximum speed (velocity) is $A\omega$ $\qquad \omega = \frac{2\pi}{T}$

maximum $E_k = \frac{1}{2} \times m \times (\text{maximum speed})^2 = \frac{1}{2} mA^2\omega^2$

$= \frac{1}{2} \times 0.1\,\text{kg} \times (0.02\,\text{m})^2 \times \left(\frac{2\pi}{2.0\,\text{s}}\right)^2 = 2 \times 10^{-4}\,\text{J}$

Always beware of non-SI units – here the mass is given in grams not kilograms. That's why it's best to write units when you are substituting into equations. Also, don't forget to square numbers when working out E_k – this is a very common calculation error.

Damping

A system that oscillates with no input or loss of energy is said to perform **free oscillations**. The frequency of these oscillations depends only on the physical properties of the system (e.g. mass and force constant). This is the system's **natural frequency** of oscillation.

If energy is lost (for example, by friction and air resistance), then the motion is said to be **damped**. The amplitude decreases with time, often exponentially, as shown in the graph in the Worked Example below.

Worked Example

The mass in the Worked Example opposite was suspended in a beaker of water and set into oscillation. A second graph was plotted. Use this graph to describe the behaviour of the mass in the water. **[4]**

Edexcel June 2000 Unit PSA6i

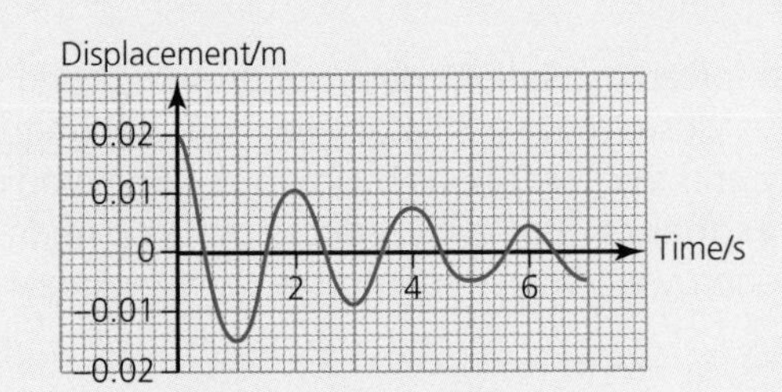

One mark each would be awarded for any four of the following:

- Resistive forces in the water…
- …dissipate energy from the oscillations…
- …and so the oscillations are damped
- The period remains constant (2.0 s),…
- …the same as it was for the undamped oscillations
- The amplitude decreases…
- …exponentially with time

Even for a question that appears rather chatty, take care to be precise in your answer with your use of the physics words. For example, it is the *amplitude* that decreases exponentially, not the *displacement*.

Quick Questions

Q1 A 200 g mass undergoes SHM with amplitude 0.4 m and period 3.0 s. Calculate:

a its maximum kinetic energy,

b the force constant k,

c the displacement when the kinetic and potential energies are equal.

Q2 A hydrogen atom vibrates in a certain molecule with frequency 8.7×10^{13} Hz. The total energy of oscillation is 5.7×10^{-20} J. Find a value for the amplitude of oscillation A. (Mass of hydrogen atom = 1.7×10^{-27} kg)

Forced oscillations and resonance

In the experiment illustrated, the mass on the spring is an oscillator. If the mass is lifted a little then released, it will oscillate freely in SHM (though with a little air resistance damping). Its SHM frequency is called its **natural frequency** (f_0).

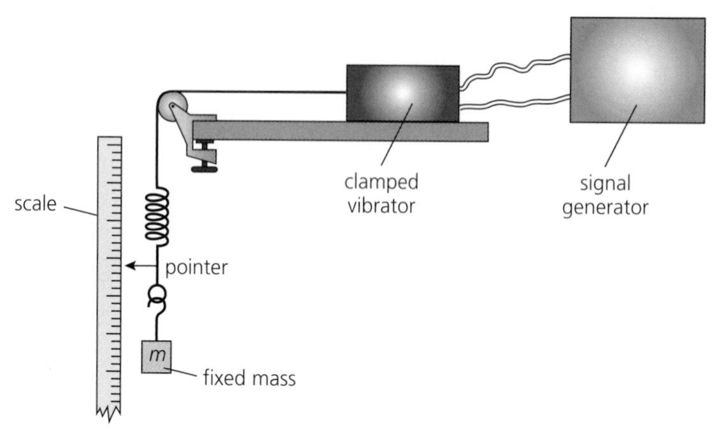

Experiment to investigate forced oscillations

The experiment is designed so that this natural oscillator can be made to undergo **forced oscillations**. The clamped vibrator moves its end of the string back and forth, forcing the spring–mass system, the natural oscillator, into motion. Energy is transferred from the forcer to the natural oscillator.

The *frequency* of the forced oscillations is the same as the frequency (f) of the vibrator doing the forcing.

The *amplitude* A of the forced oscillations depends on the driving frequency f. When the driving frequency f is close to the oscillator's own natural frequency f_0, there is a large transfer of energy and the oscillations build up to large amplitude. This special situation is known as **resonance**. Damping reduces the amplitude of the forced oscillations, and more energy is transferred into the surroundings.

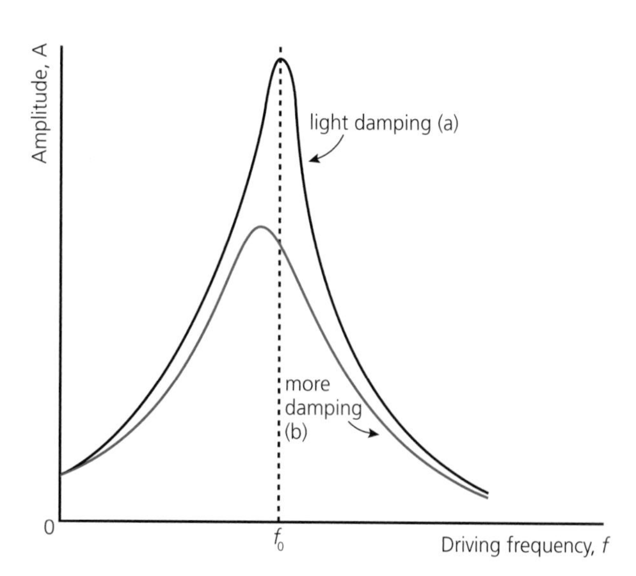

Response of a driven oscillator and resonance with (a) light damping and (b) more damping

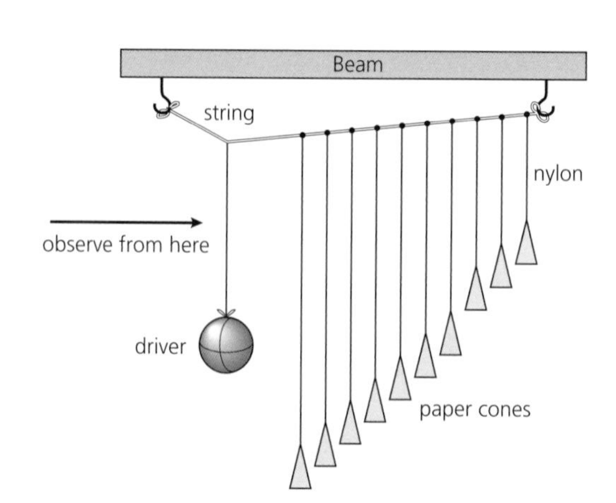

Barton's pendulums

Barton's pendulums illustrate resonance. Only the paper pendulum that is the same length as the driver achieves a large amplitude and resonance, since it is the only one with the natural frequency the same as the driver.

Examples of situations where resonance is a *desirable* effect are:

- pushing a child on a swing
- a tuned radio circuit responding to a selected frequency
- a musical instrument playing a particular note.

Materials designed especially for sound-proofing make use of both resonance and damping:

- tiny partcles in the material oscillate at sound frequencies
- the particles absorb the sound energy as they resonate
- material surrounding the particles deforms plastically, damping the oscillations
- so the energy of the sound is converted to heat in the material.

Examples of situations where resonance is *undesirable* are:

- a bridge set into oscillations by wind or humans (e.g. Tacoma Narrows, Millennium Bridge)
- earthquakes causing buildings to vibrate at their natural frequency
- machinery parts resonating at particular rates of revolution (e.g. washing machine panels)
- vehicle suspension systems resonating as the vehicle crosses rough ground.

To combat unwanted resonance in structures, designers have two possibilities:

- add more damping, for example by using shock absorbers which introduce friction and dissipate energy as heat
- change the natural frequency of the structure, either by making it stiffer or floppier (changed k so changed f_0), or by adding or removing mass (changed m so changed f_0).

Quick Questions

Q1 A car driver notices that, when she drives at a particular speed, the rear-view mirror shakes visibly. In an attempt to stop it, she sticks a large lump of Blu-tack to the back of the mirror.
a Explain why the vibration is noticeable only at a particular speed.
b Explain how the Blu-tack will affect the vibrations.

Q2 **a** Why is it desirable to have the lightest possible damping in a tuned radio circuit?
b How would that be achieved in a circuit?

Q3 **a** Why is machinery usually designed to rotate at a frequency well above the natural oscillation frequency of any of the parts of the structure?
b Why may brief resonance of a part be noticed when the machinery is first switched on?

Thinking Task

Look back at the diagram of the experiment at the top of the opposite page. This system actually resonates at several different frequencies – not just at the natural frequency of the mass on the end of the spring. Can you suggest other possible oscillations?

Section 6: Oscillations checklist

By the end of this section you should be able to:

Revision spread	Checkpoints	Spec. point	Revised	Practice exam questions
Simple harmonic motion (SHM)	Recall that the condition for simple harmonic motion is $F = -kx$. Identify situations in which simple harmonic motion will occur.	119		
	Recognise and use the expressions $a = -\omega^2 x$, $a = -A\omega^2 \cos \omega t$, $v = -A\omega \sin \omega t$, $x = A \cos \omega t$, and $T = \frac{1}{f} = \frac{2\pi}{\omega}$ as applied to a simple harmonic oscillator.	120		
	Obtain a displacement–time graph for an oscillating object and recognise that the gradient at a point gives the velocity at that point.	121		
Energy and damping in SHM	Recall that the total energy of an undamped simple harmonic system remains constant and recognise and use expressions for the total energy of an oscillator.	122		
	Distinguish between free and damped oscillations.	123		
Forced oscillations and resonance	Distinguish between free, damped and forced oscillations.	123		
	Investigate and recall how the amplitude of a forced oscillation changes at and around the natural frequency of a system. Describe qualitatively how damping affects resonance.	124		
	Explain how damping and the plastic deformation of ductile materials reduce the amplitude of oscillation.	125		

ResultsPlus

Build Better Answers

A spring of negligible mass and spring constant k has a load of mass m suspended from it. A student displaces the mass and releases it so that it oscillates vertically.

The student investigates the variation of the time period of the vertical oscillations T with m. Describe how he could verify experimentally that $T \propto \sqrt{m}$. Include any precautions the student should take to make his measurements as accurate as possible. You may be awarded a mark for the clarity of your answer.

[5]

Edexcel June 2006 Unit Test 4

Examiner tip

Any question where you have to describe an experiment requires three key elements in its answer:

1 What (additional) apparatus will you use?
2 What will you measure?
3 What will you do with the results?

There will also be other details to include which are specific to the problem set.

Student answer	Examiner comments
• Time T with a stopwatch. • Start and stop the watch as m passes a reference mark in the middle. • Repeat the timing several times and take an average. • Use several different values of mass. • Use a computer program to show required result.	This answer starts well, but lacks detail in its later stages. ■ A **basic answer** would be as given here, and would gain 2 marks. ▲ An **excellent answer** would include the following points: • That for accurate results you will time at least ten oscillations in one go, rather than repeatedly timing just one oscillation. • The number of different values of mass you will use – five is usually the minimum needed. • The *range* of masses, for example 100 g, 200 g, 300 g, 400 g, 500 g. • An *explanation* of what the computer would be used for. • That you will plot a graph of T against $\sqrt{m}$, and that this should be straight and through the origin to verify $T \propto \sqrt{m}$.

Practice exam questions

1 The graph shows how the displacement of a child on a playground swing varies with time.

The maximum velocity of the child is

A $\frac{\pi}{2}\,\mathrm{m\,s^{-1}}$

B $\pi\,\mathrm{m\,s^{-1}}$

C $2\pi\,\mathrm{m\,s^{-1}}$

D $3\pi\,\mathrm{m\,s^{-1}}$ **[1]**

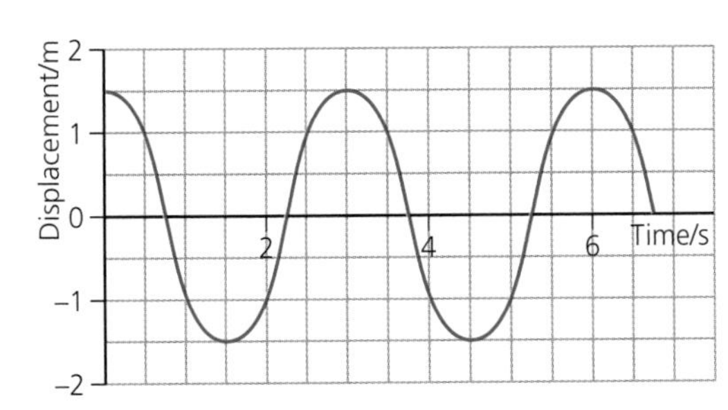

2 In the diagram a mass m is swinging backwards and forwards on the end of a string, along the path G–H–J–K–J–H ... etc. Which of these statements is true?

A The speed of m is greatest at G and K.

B The kinetic energy of m is always increasing as it passes through J.

C The total energy of m increases as it moves from G to H.

D m is instantaneously stationary at G and K. **[1]**

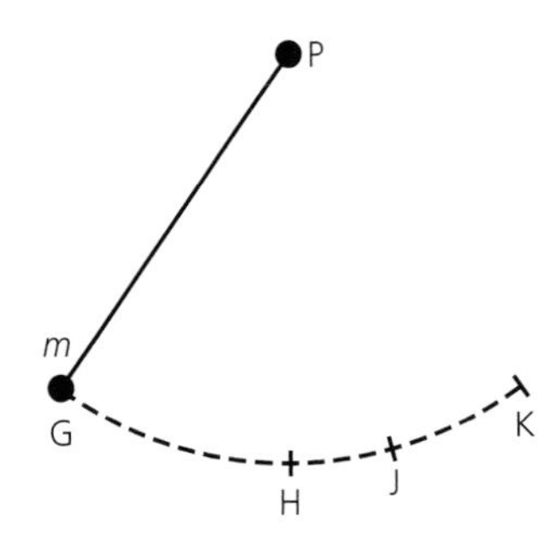

3 When a person walks across a suspended footbridge, the bridge can oscillate with increasing amplitude.

a Name the effect which causes this and state the condition needed for the amplitude to increase in this situation. **[2]**

b In November 1940, the wind caused some alarming movement and twisting of the road bridge over Tacoma Narrows in the United States. The amplitude of the oscillations became so large that cars were abandoned on the bridge.

i Why can these oscillations be described as forced? **[1]**

ii The vertical oscillations of the bridge can be modelled using the equations of SHM. Calculate the maximum acceleration of the bridge when it was oscillating 38 times per minute and the amplitude of its oscillations was 0.90 m. **[2]**

iii Use this value to explain why any car abandoned on the bridge would lose contact with the road's surface at a certain point in the oscillation. Identify this point. **[2]**

Edexcel June 2008 Unit PSA5

Gravity and orbits

Two masses m_1 and m_2 attract each other with a force F, which is described by an inverse-square law:

$$F = \frac{Gm_1m_2}{r^2} \quad \text{or} \quad \frac{GMm}{r^2}$$

where r is the distance apart of their centres, and G is the universal gravitational constant ($G = 6.67 \times 10^{-11}\,\text{N m}^2\,\text{kg}^{-2}$). This is known as **Newton's law of gravitation**.

Gravitational field strength

The strength g of a **gravitational field** is defined as the force acting on unit mass placed at a point in the field. Therefore, from Newton's law of gravitation, the field around mass m_1 can be derived by treating m_2 as the unit mass:

$$g = \frac{F}{m} = -\frac{GM}{r^2}$$

The minus sign is inserted to indicate that this is an **attracting** force. If r is measured outwards from m_1, then the field is in the opposite direction. At the Earth's surface $g = 9.81\,\text{N kg}^{-1}$.

ResultsPlus
Watch out!

When pressing your calculator buttons, don't forget to square the radius.

Worked Example

Using the equation above, calculate a value for the mass of the Earth. (Radius of Earth = 6.38×10^6 m)

$$g = (-)\frac{GM_1}{r^2}$$

$$\therefore M_1 = \frac{gr^2}{G} = \frac{9.81\,\text{N kg}^{-1} \times (6.38 \times 10^6\,\text{m})^2}{6.67 \times 10^{-11}\,\text{N m}^2\,\text{kg}^{-2}} = 5.99 \times 10^{24}\,\text{kg}$$

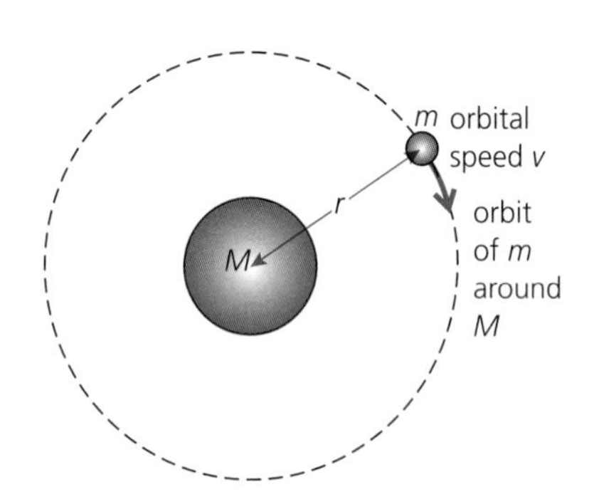

Circular orbits

If an object of mass m is in circular orbit round another much bigger mass, M, then the gravitational force provides the necessary centripetal force:

$$F = \frac{mv^2}{r} = \frac{GMm}{r^2}$$

$$\text{so } v^2 = \frac{GM}{r}$$

(Note that m cancels out at this stage.)

The orbital speed v is related to the orbit period T:

$$v = \frac{2\pi r}{T}$$

These relationships can be combined to relate the period of an orbit to its radius, e.g. for planets orbiting the Sun, or satellites orbiting a planet:

$$T^2 = \frac{4\pi^2 r^3}{GM}$$

Worked Example

A geostationary satellite orbits so that it is always over the same point on the Earth.

a Calculate its height above the Earth's surface.

b Explain why it must always be over a point on the equator.

---

a Work out its orbit radius. Use the data for M in the Worked Example opposite.

$$T^2 = \frac{4\pi^2 r^3}{GM}$$

$$\therefore r^3 = \frac{GMT^3}{4\pi^2} = \frac{6.67 \times 10^{-11}\,\text{N m}^2\,\text{kg}^{-2} \times 5.99 \times 10^{24}\,\text{kg} \times (24 \times 60 \times 60\,\text{s})^2}{4\pi^2}$$

$= 7.55 \times 10^{22}\,\text{m}^3$ and $r = 4.23 \times 10^7\,\text{m}$

and $h = 4.23 \times 10^7\,\text{m} - 6.38 \times 10^6\,\text{m} = 3.59 \times 10^7\,\text{m}$

b A satellite has to orbit with the centre of the Earth as the centre of its circle. So if it wasn't over the equator it would spend half its time over the northern hemisphere, and half its time over the southern hemisphere – so not always over the same point.

Don't forget that r is the distance from the satellite to the centre of the Earth. So to get the height of the orbit, you must subtract the Earth's radius.

Comparing electric and gravitational fields

	Electric	Gravitational
acts on	charge	mass
types of charge/mass	+ and –	just mass!
direction of forces	like charges repel (+ve force) unlike charges attract (–ve force)	always attracting (–ve)
field strength	$E = \frac{F}{q}$	$g = \frac{F}{m}$
force equation	$F = \frac{kQ_1Q_2}{r^2}$	$F = -\frac{Gm_1m_2}{r^2}$
field due to point mass/charge	$\frac{kQ}{r^2}$	$-\frac{Gm}{r^2}$

Quick Questions

Q1 The Moon's orbital period is 27.3 days.
- **a** Using the data on these pages, calculate its radius of orbit.
- **b** How many Earth's radii is this orbit radius (to the nearest whole number)?
- **c** Calculate the Earth's gravitational field at this distance.
- **d** Explain why this is also the centripetal acceleration of the Moon.

Q2 For each of the following statements, say whether it applies to **A** gravitational fields, **B** electric fields, or **C** both.
- **a** The field causes a force that depends only on an object's charge.
- **b** The field causes a force that depends only on an object's mass.
- **c** Spherical objects have a field whose strength obeys an inverse-square law.
- **d** The field strength is close to zero a long way away from the object.
- **e** The force due to the field is always attractive.
- **f** The force can be attracting or repelling.

Thinking Task

The gravitational field strength at the Earth's surface is g. Which of the following is the gravitational field strength at the surface of a planet which is half as dense as the Earth and whose diameter is four times that of the Earth?

A $2g$ **B** $4g$ **C** g

D $\frac{g}{2}$ **E** $8g$

This sort of question needs a lot of care!

Stars

The **flux** F (or intensity) of radiation from a star is the power received per unit area of detector on Earth. If the distance d to a star is known then its **luminosity** L can be calculated from its flux using an inverse-square law:

$$F = \frac{L}{4\pi d^2}$$

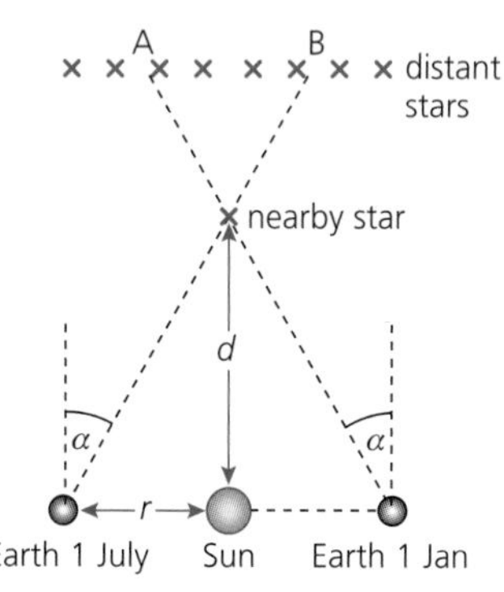

Finding distance to a star using parallax

A star's luminosity is the total power it emits. It is usually expressed as a multiple of the Sun's luminosity, L_{Sun}.

Two possible ways of finding the distance d to a star are:

- using a **parallax method**. Observe the star from opposite sides of the Earth's orbit around the Sun, and make careful measurements of its apparent position. Then calculate its distance using trigonometry.
- using a '**standard candle**'. The luminosity of some stars can be found from other measurements – perhaps the particular spectrum of radiation they emit, or a regular variation in their brightness which can be matched to that of known stars. We can always measure the flux F from a star; so knowing F and L we can calculate d.

Worked Example

At the top of the Earth's atmosphere the intensity F of the Sun's radiation is about $1.4\,\text{kW m}^{-2}$. The distance from Earth to Sun is 1.50×10^{11} m. Calculate the Sun's luminosity.

$$F = \frac{L}{4\pi d^2} \text{ so } L = 4\pi d^2 F = 4\pi \times (1.50 \times 10^{11}\,\text{m})^2 \times 1.4\,\text{kW m}^{-2}$$

$$= 4.0 \times 10^{23}\,\text{kW}$$

Laws about radiation

The radiation from a star follows the laws of 'black body' radiation.

1 Stefan-Boltzmann law:

$$L = \sigma T^4 \times \text{surface area}$$

For a sphere, $L = 4\pi r^2 \sigma T^4$

where T is the kelvin temperature of the star's surface. σ is called the Stefan-Boltzmann constant: $\sigma = 5.67 \times 10^{-8}\,\text{W m}^2\,\text{K}^{-4}$.

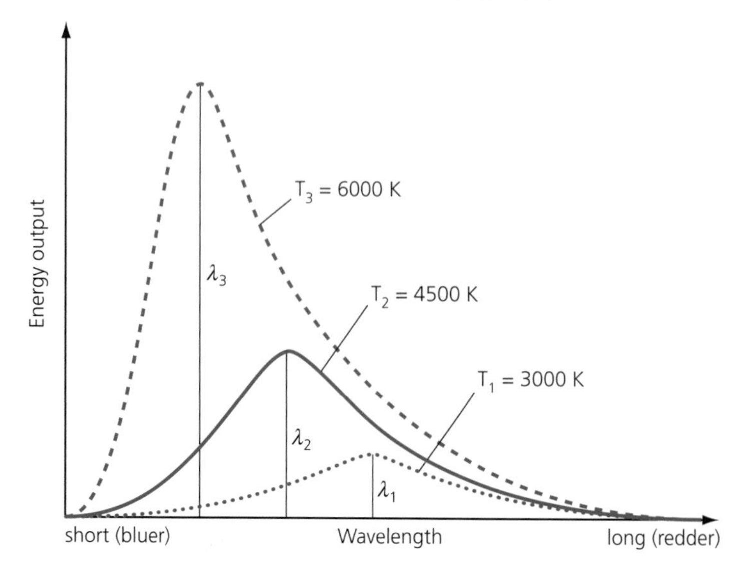

Radiation curves for different temperatures

2 Wien's law:

$$\lambda_{max} T = \text{constant } (2.898 \times 10^{-3}\,\text{m K})$$

where λ_{max} is the wavelength at which the energy output peaks.

Worked Example

Using the equations above, calculate the wavelength at which the Sun emits maximum energy. Radius of the Sun = 6.96×10^8 m.

$\lambda_{max} T = 2.898 \times 10^{-3}\,\text{m K}$

$$\therefore \lambda_{max} = \frac{2.898 \times 10^{-3}\,\text{m K}}{5800\,\text{K}} = 5.00 \times 10^{-7}\,\text{m}$$

Properties of stars

The properties of many stars can be summarised using a **Hertzsprung-Russell (HR) diagram**. The two key properties of a star – luminosity and surface temperature – place it at a particular point on the diagram.

Note that both axes on an HR diagram have logarithmic scales, and that temperature increases to the left. The star marked A is very hot and very luminous.

Stars form from huge clouds of gas and dust. They join the **main sequence** soon after they are formed, and settle to an almost constant luminosity and temperature while hydrogen fusion takes place. When all the hydrogen in the core has turned to helium, the star swells, cools, and becomes a red giant. After fusion of heavier nuclei stops, the star cools and shrinks to become a white dwarf.

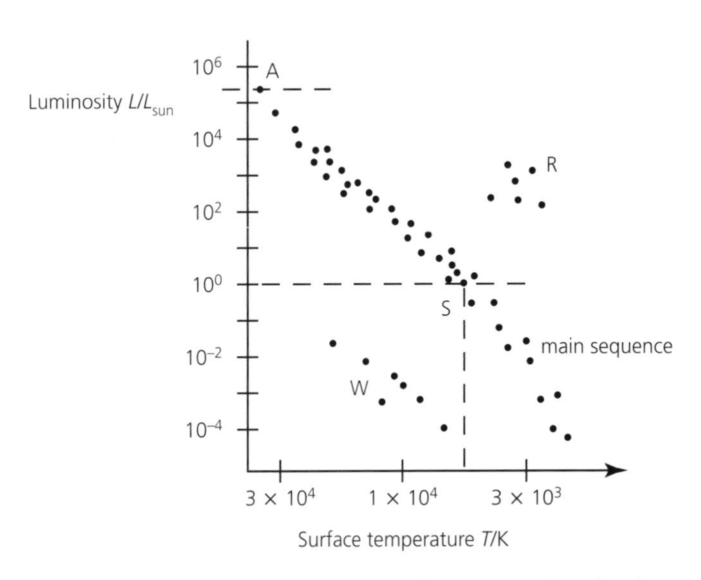

Hertzsprung-Russell diagram: S = Sun, R = red giant, W = white dwarf

If the initial mass of the star is greater than about eight times that of our Sun, this shrinkage can be so rapid that a **supernova** explosion occurs.

Be particularly careful when you have to read off values from a log scale. The intervals are not all the same!

Quick Questions

Q1 A star has a luminosity $L = 1.9 \times 10^{25}$ W and radius $r = 4.2 \times 10^8$ m. Calculate:
- **a** the surface temperature of the star,
- **b** the wavelength at which the star emits maximum energy.

Q2
- **a** The star in Q1 is 11.6 light years from Earth. Calculate the flux of radiation received from it at the top of the atmosphere. (1 light year = 9.5×10^{15} m)
- **b** How might the appearance of the star differ from that of the Sun to someone looking at both from another galaxy?

Q3 From the HR diagram, estimate the luminosity of the star labelled A. Give your answer as a multiple of the Sun's luminosity.

The universe

Light from stars and galaxies contains **emission lines** and **absorption lines** – certain wavelengths where the light is much brighter or fainter than at other wavelengths. The lines depend on which atoms, molecules and ions are present. Nearby galaxies all have a very similar pattern of lines.

Redshift

Light from distant galaxies contains lines whose wavelengths are stretched compared with those of nearby galaxies. The stretch, or **redshift**, z, is related to a galaxy's speed v away from us. Provided v is much less than the speed of light c,

$$z = \frac{\Delta\lambda}{\lambda} \approx \frac{v}{c}$$

where λ is the original wavelength and $\Delta\lambda$ is the difference between the original wavelength and the wavelength observed.

Measurements of redshift and the distance of the galaxy from us show that the greater the distance, the greater the redshift. Galaxies are moving apart, and the universe is expanding.

Hubble's law

The proportional relationship between recession speed v and distance d is known as **Hubble's law**. The gradient of the graph gives the **Hubble constant** H_0.

$$v = H_0 d$$

Astronomical distances are commonly expressed in **parsecs** (**pc**) or megaparsecs (Mpc), and recession speeds in km s^{-1}; so H_0 has units $\text{km s}^{-1}\,\text{Mpc}^{-1}$. $1\text{ pc} = 3.09 \times 10^{16}\text{ m}$.

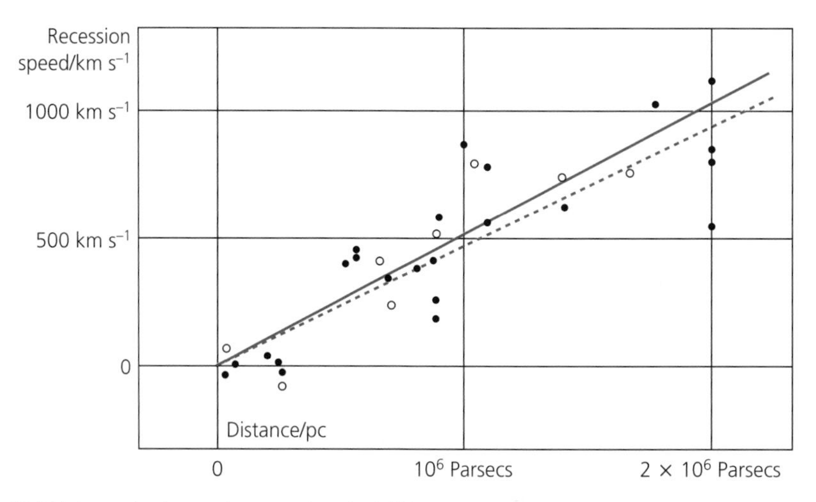

Hubble's graph of recession speed against distance

Worked Example

Light from a distant galaxy has an emission line with wavelength 445 nm. The line has wavelength 410 nm when measured in the lab. Using $H_0 = 71\ \text{km s}^{-1}\text{Mpc}^{-1}$, calculate **a** the galaxy's red shift, **b** its recession speed, **c** its distance.

a $\Delta\lambda = 445\ \text{nm} - 410\ \text{nm} = 35\ \text{nm}$

$$z = \frac{\Delta\lambda}{\lambda} = \frac{35\ \text{mm}}{410\ \text{nm}} = 0.0854$$

b $v = cz = 0.0854 \times 3.00 \times 10^8\ \text{m s}^{-1} = 2.56 \times 10^7\ \text{m s}^{-1} = 2.56 \times 10^4\ \text{km s}^{-1}$

c $d = \frac{v}{H_0} = \frac{2.56 \times 10^4\ \text{km s}^{-1}}{71\ \text{km s}^{-1}\ \text{Mpc}^{-1}} = 360\ \text{Mpc}$

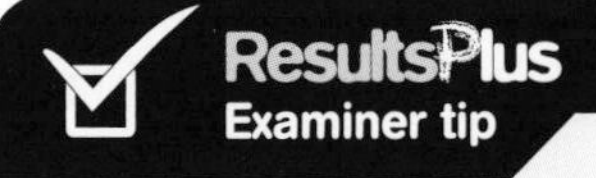

Keep careful track of the units here. Convert v into km s^{-1} so that it cancels with the km s^{-1} in H_0. You need not worry about converting Mpc to SI units, just leave the answer in Mpc.

Expansion of the universe

From Hubble's law we can deduce the time it has taken for galaxies to get where they are now. This is the time since the beginning of the expansion – the beginning of the universe, known as the Big Bang. If we assume that the universe has always been expanding at its current rate, then

$$\text{age of universe} = \frac{d}{v} = \frac{1}{H_0}$$

If the universe is dominated by only gravitational forces, then since the Big Bang these will have slowed the expansion. The future of the universe then depends on the amount of mass in the universe.

- An **open universe** (least mass) will keep on expanding for ever, even though the expansion will go on slowing down.
- A **flat** or **critical universe** (more mass) will stop expanding, but only at an infinite time in the future.
- A **closed universe** (most mass) will stop expanding and start to contract again.

Most estimates of the mass of the universe suggest that we are close to the critical situation. Only about 10% of matter is visible in the form of stars and gas clouds. The other 90% is called **dark matter**, whose nature is still unknown. The existence of dark matter is deduced from measuring the movements of stars and galaxies.

There is recently discovered evidence that the expansion of the universe is actually speeding up. To account for this scientists have suggested something at work called 'dark energy' – but what this might be also remains a mystery at the moment.

Quick Questions

Q1 An emission line has wavelength 434 nm in the lab. The same line has wavelength 512 nm when observed in the light of a distant galaxy. Assuming $H_0 = 71\ \text{km s}^{-1}\ \text{Mpc}^{-1}$, calculate the distance to the galaxy.

Q2 **a** Express $H_0 = 71\ \text{m s}^{-1}\ \text{Mpc}^{-1}$ in units of s^{-1}.

b Hence estimate the age of the universe in seconds and in years. ($1\ \text{pc} = 3.09 \times 10^{16}\ \text{m}$; $1\ \text{year} = 3.16 \times 10^7\ \text{s}$)

c Explain why this is an estimate.

Thinking Task

a Write a short paragraph explaining the meaning of the phrase 'closed universe'.

b Explain how astronomical observations in the very distant future in a closed universe would differ from those of today.

Section 7: Astrophysics and cosmology checklist

By the end of this section you should be able to:

Revision spread	Checkpoints	Spec. point	Revised	Practice exam questions
Gravity and orbits	Use the expression $F = \frac{Gm_1m_2}{r^2}$.	126		
	Derive and use the expression $g = \frac{-Gm}{r^2}$ for the gravitational field due to a point mass.	127		
	Recall similarities and differences between electric and gravitational fields.	128		
Stars	Recognise and use the expression relating flux, luminosity and distance, $F = \frac{L}{4\pi d^2}$ and understand its application to standard candles.	129		
	Explain how distances can be determined using trigonometric parallax.	130		
	Explain how distances can be determined by measurements on radiation flux received from objects of known luminosity (standard candles).	130		
	Recognise and use a simple Hertzsprung-Russell diagram to relate luminosity and temperature, and to explain the life cycle of stars.	131		
	Recognise and use the expression $L = \sigma T^4 \times$ surface area, (for a sphere $L = 4\pi r^2 \sigma T^4$) (Stefan-Boltzmann law) for black body radiators.	132		
	Recognise and use the expression $\lambda_{max} T = 2.898 \times 10^{-3}$ m K (Wien's law) for black body radiators.	133		
The universe	Recognise and use the expressions $z = \frac{\Delta\lambda}{\lambda} \approx \frac{v}{c}$ for a source of electromagnetic radiation moving relative to an observer.	134		
	Recognise and use the expression $v = H_0 d$ for objects at cosmological distances.	134		
	Be aware of the controversy over the age and ultimate fate of the universe associated with the value of the Hubble constant and the possible existence of dark matter.	135		

ResultsPlus
Build Better Answers

Discuss the ultimate fate of the universe. Your answer should include reference to dark matter and reasons why the fate of the universe is uncertain. **[6]**

Edexcel June 2001 Unit Test 4

Examiner tip

- For this sort of open-ended question, pause and plan your answer before you start writing.
- Look at the number of marks. Here there are 6 marks available, so you should aim to make at least six points that are *relevant* and *good physics*.
- Your answer can be quite short and still gain full marks. A long rambling answer will not necessarily score high marks. A good answer should be able to fit comfortably into the space provided on the exam paper.
- The examiners' mark scheme will contain more points than the marks available, so there are many different answers that can gain full marks.
- The points listed below are those that the examiners might expect to see. Other correct relevant points would also gain marks.

Student answer

Possible student answers

- The universe may continue to expand…
- …or may collapse back on itself
- The future depends on the amount of matter in the universe…
- ...because mass leads to gravitational forces/deceleration of galaxies
- So far, not enough ordinary matter has been identified to halt the expansion…
- there is also dark matter in the universe…
- …e.g. WIMPs, black holes, neutrinos,…
- …which cannot be detected by observations of radiation…
- …but it exerts gravitational force
- Recent evidence suggests that the expansion is actually accelerating…
- …driven by 'dark energy'…
- …which is not understood.

Practice exam questions

1 The gravitational field strength on the surface of the Earth is g. The gravitational field strength on the surface of a planet of twice the radius and the same density is

A $4g$ **B** $2g$ **C** g **D** $\frac{g}{4}$ **[1]**

Edexcel sample assessment material 2007 Unit Test 5

2 Cosmic background is a remnant of the Big Bang and appears to pervade the universe. It has a maximum wavelength in the microwave region of the electromagnetic spectrum. This can be calculated to correspond to a temperature of about 3 K. This calculation is based on the assumption that

A the universe is spherical
B the universe is expanding
C space can be regarded as a black body
D space is a vacuum. **[1]**

Edexcel sample assessment material 2007 Unit Test 5

3 The x-axis of a Hertzsprung-Russell diagram is log T. This is because

A the range of temperatures of the surfaces of stars is large
B the temperatures of the surfaces of stars are all very large numbers
C the scale has to start with the hottest stars
D the diagram would be impossible to interpret if log T was the y-axis. **[1]**

Edexcel sample assessment material 2007 Unit Test 5

4 The graph shows how the logarithm of the electrical power P supplied to a filament lamp varies with the logarithm of the temperature T of the filament.

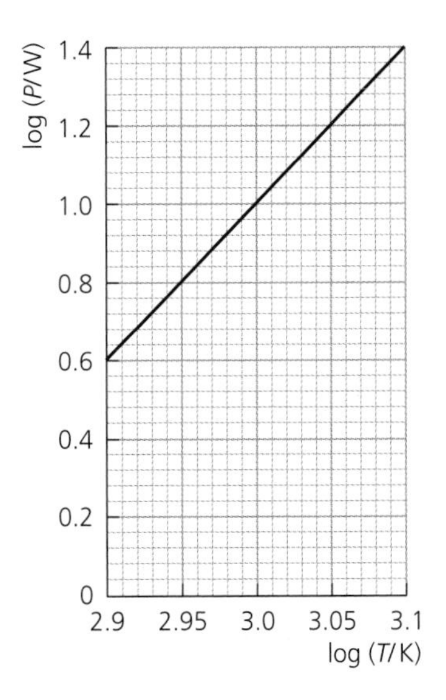

a P is related to T by a power law: $P = kT^n$. Use the graph to determine n. **[2]**

b A student suggests that this relationship is predicted by the Stefan-Boltzmann law. Comment on this statement. **[2]**

Edexcel sample assessment material 2007 Unit Test 5

5 The Hubble constant is thought to be about 70 000 m s^{-1} Mpc^{-1}.

a Give one reason why the value of this constant is uncertain. **[1]**

b State how an estimate of the age of the universe can be calculated from the Hubble constant. **[1]**

c Explain how the ultimate fate of the universe is associated with the Hubble constant. **[3]**

Edexcel sample assessment material 2007 Unit Test 5

Unit 5: Practice unit test

Section A

1 A certain mass of air is contained in a cylinder. Which of these changes would increase the density of the air?
A heating the air, keeping its volume the same
B increasing the volume of the air, keeping its temperature the same
C heating the air, keeping its pressure the same
D increasing the pressure of the air, keeping its temperature the same **[1]**

2 Which of these could be spontaneously emitted by a radioactive source and penetrate several centimetres of lead shielding, but would *not* be affected by a magnetic field?
A α particle **B** γ radiation **C** proton **D** neutron **[1]**

3 The spectrum of visible light from the Sun contains a number of dark lines known as Fraunhofer lines.

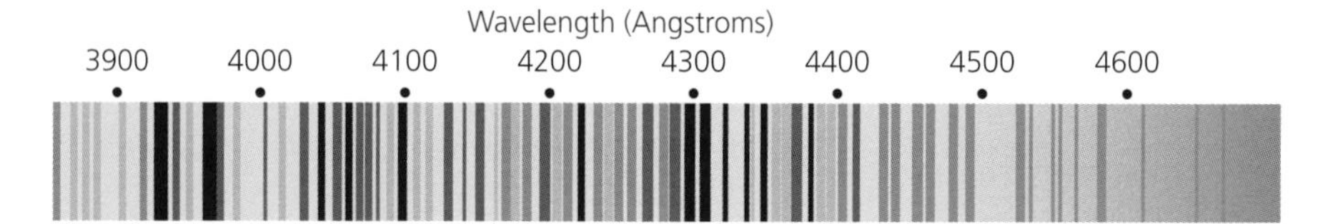

This is due to
A polarisation caused by the atmosphere
B absorption of light by atoms
C refraction of light by telescopic lenses
D the Doppler effect because atoms are moving very fast. **[1]**

Edexcel sample assessment material 2007 Unit Test 5

4 The rise and fall of the tides on a beach follows approximate SHM. There are two high tides every 24 hours. On one day there is a high tide at midday. At about what time after that is the tide running out at the fastest rate?
A 1.30 pm **B** 3.00 pm **C** 6.00 pm **D** 9.00 pm **[1]**

5 A liquid X has a specific heat capacity half that of water. The same amount of heat is supplied to 3 kg of X and to 1 kg of water. The temperature rise of X is
A $\frac{1}{6}$ times that of the water
B $\frac{2}{3}$ times that of the water
C $\frac{3}{2}$ times that of the water
D 6 times that of the water. **[1]**

6 The half life of a radioactive substance can be doubled by
A halving the mass of the sample
B doubling the kelvin temperature of the sample
C doubling the pressure on the substance
D none of the procedures in choices A–C. **[1]**

7 One end of a spring XYZ is attached to the ceiling. The other end carries a mass m which oscillates up and down.

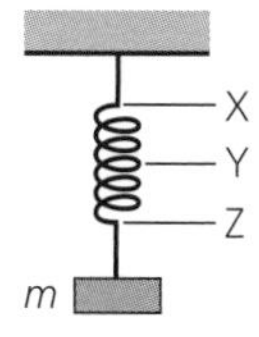

Which of the changes shown and described below would *increase* the frequency of these oscillations?

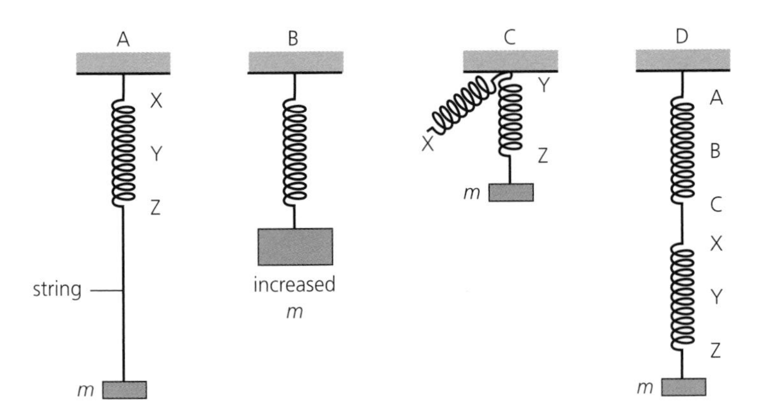

A attaching a long piece of inelastic string between Z and m
B increasing the mass m
C attaching the point Y to the ceiling instead of the point X
D adding a second identical spring ABC in series with XYZ **[1]**

8 All quasars show large red shifts in the light received from them. This shows that all quasars
A have large diameters
B are moving towards us at very high speeds
C have a low temperature
D are far away from Earth **[1]**

Edexcel sample assessment material 2007 Unit Test 5
[8 marks]

Section B

9 An upright kitchen freezer – a heat pump – has an internal volume of $0.20\,m^3$ and maintains a steady internal temperature of $-18\,°C$. After the door has been left open for a short time all the cold air in the freezer has been replaced by air at a room temperature of $22\,°C$.

Show that the energy that needs to be extracted from this air to cool it down to freezer temperature is about 6 kJ. Take the density of air to be $1.3\,kg\,m^{-3}$ and the specific heat capacity of air to be $610\,J\,kg^{-1}\,K^{-1}$. **[4]**

Edexcel June 2008 Unit Test 6

10 a A planet of mass m orbits a star of mass M. The radius of orbit is r. By considering the force required for circular motion in this situation, show that the period T of the orbit is given by

$$T^2 = \frac{4\pi^2 r^3}{GM}$$ **[3]**

b Measurements have shown that star HD70642 has a planet which orbits the star with a period of about 6 years. The radius of the orbit is about 3 × the radius of the Earth's orbit around the Sun.

i Use the formula in **a** to find a value for the ratio $\frac{\text{mass of star HD70642}}{\text{mass of Sun}}$. **[3]**

ii Because of the presence of the planet, the star HD70642 does not remain at rest. Instead, the planet and star both orbit their common centre of mass. Explain why the orbiting speed of the star is very small in comparison to the speed of the planet. **[2]**

c Astronomers discovered the planet by observing the 'Doppler Wobble' effect. As the planet orbits the star, light from the star undergoes a Doppler shift in its frequency. Explain why this method is likely to detect only *very large* planets. **[3]**

Edexcel sample assessment material 2007 Unit Test 5

11 A satellite uses a radium-226 source as a back-up power supply. Radium-226 is an alpha-particle emitter.

a The satellite requires a back-up power of 55 W. Each alpha particle is emitted with an energy of 7.65×10^{-13} J. Show that the activity of the source must be about 7×10^{13} Bq. **[2]**

b Radium-226 has a half-life of 1620 years. Show that its decay constant is about $1.4 \times 10^{-11}\,s^{-1}$. 1 year = 3.15×10^{7} s. **[2]**

c Hence determine the number of radium-226 nuclei that would produce the required activity. **[2]**

d Calculate the mass of radium-226 that would produce a power of 55 W. 226 g of radium-226 contains 6.02×10^{23} nuclei. **[2]**

e In practice this mass of radium-226 produces more than 55 W of power. Suggest a reason why. **[1]**

Edexcel June 2008 Unit Test 1

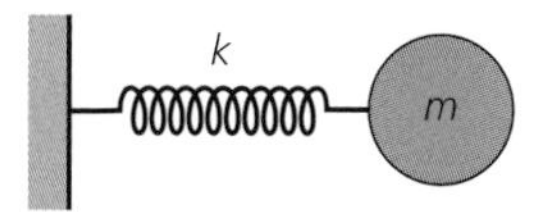

12 Certain molecules such as hydrogen chloride (HCl) can vibrate by compressing and extending the bond between atoms. A simplified model ignores the vibration of the chlorine atom and just considers the hydrogen atom as a mass m on a spring of stiffness k which is fixed at the other end.

a i Show that the acceleration of the hydrogen atom, a, is given by $a = \frac{-kx}{m}$ where x is the displacement of the hydrogen atom. **[2]**

ii Hence derive the equation $T = 2\pi\sqrt{\frac{m}{k}}$ for the period of natural oscillations of the hydrogen atom. **[2]**

b Infrared radiation is used in chemical analysis. Compared to other radiations, infrared radiation of wavelength 3.3 μm is strongly absorbed by hydrogen chloride gas. As a result of this absorption, the amplitude of oscillations of the hydrogen atoms significantly increases.

i What name is given to this phenomenon? **[1]**

ii State the condition for it to occur. **[1]**

iii Calculate the frequency of infrared radiation of wavelength 3.3 μm. **[2]**

iv Hence calculate the stiffness of the hydrogen chloride bond. Mass of hydrogen atom = 1.67×10^{-27} kg. **[3]**

Edexcel sample assessment material 2007 Unit Test 5

13 A problem with warming milk in a saucepan is that it can suddenly boil over if it is not watched carefully. A student decides to take some measurements to find the time it takes for the milk to reach a temperature of 96 °C so he can be ready for it without having to keep a constant watch.

a The student first uses an electric hotplate to warm a saucepan of water from room temperature to 96 °C. He measures the time taken to be 347 s. He calculates the heat energy gained by the water to be 1.63×10^{5} J. Show that the rate at which heat energy is supplied to the water by the electric hotplate is about 500 W. **[2]**

b The student then uses the following data to calculate the time taken for milk taken from a refrigerator to reach the temperature of 96 °C.

mass of milk = 0.44 kg
initial temperature of milk = 12 °C
desired final temperature of milk = 96 °C
specific heat capacity of milk = 3800 $J\,kg^{-1}\,°C^{-1}$

i Show that the heat energy the milk needs to gain is about 1×10^{5} J. **[2]**

ii Calculate the time it would take for the milk to reach the temperature of 96 °C. Assume that the student uses the same hotplate as in **a**. **[2]**

c The student warms up the milk and is surprised when the time taken is exactly the time calculated. He had expected it to take longer because of heat losses.

i Explain why he might expect it to take longer. **[1]**

ii Suggest why the calculated time was the same as the actual time. **[1]**

Edexcel June 2008 Unit PSA1

14 Read the following passage and answer the questions that follow.

A nova is a sudden brightening of a star. Novae are thought to occur on the surface of a white dwarf star which is paired with another star in a binary system. If these two stars are close enough to each other, hydrogen can be pulled from the surface of the star onto the white dwarf. Occasionally, the temperature of this new material on the surface of the white dwarf may become hot enough for the hydrogen to fuse to helium. This causes the white dwarf to suddenly become very bright. In a nova, this hydrogen fusion occurs by the 'CNO' process, where helium-4 is produced by a series of steps in which protons react with various isotopes of carbon, nitrogen and oxygen. Novae are used by astronomers as standard candles.

a Complete the equation which shows a typical part of the CNO process.

$^{\cdots}_{8}O + ^{\cdots}_{\cdots}H \rightarrow ^{14}_{\cdots}N + ^{\cdots}_{\cdots} \ldots$ **[3]**

b What is a white dwarf? Suggest why hydrogen fusion in the white dwarf is likely to be the CNO process. **[3]**

c The temperature required for these processes is 10^7 K.

i Calculate the mean kinetic energy, in keV, of the particles involved. **[3]**

ii Explain how this temperature arises. **[2]**

d Astronomers use novae as standard candles. Explain what a standard candle is, and suggest what this implies about the processes occurring in a nova. **[3]**

Edexcel sample assessment material 2007 Unit Test 5

15 Sea waves are being studied using a buoy anchored in a harbour. As the waves pass the buoy they make it perform simple harmonic motion in the vertical direction. A sensor inside the buoy measures its acceleration. The graph on the right shows how this acceleration varies with time.

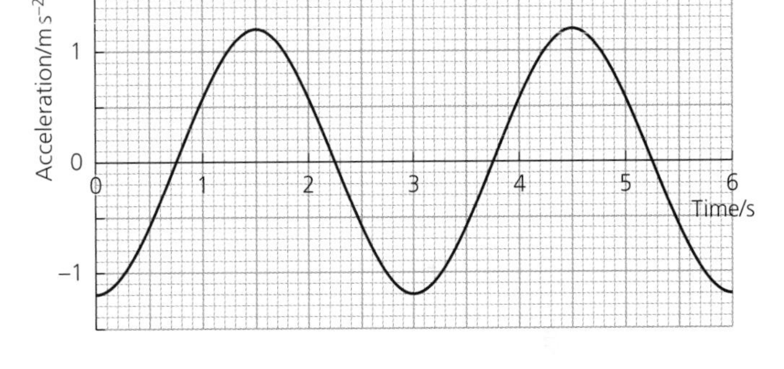

a State values for the period and maximum acceleration of the buoy. **[1]**

b Calculate the amplitude of oscillation of the buoy. **[3]**

c Sketch a graph of the displacement of the buoy against time, over the same interval of time as the acceleration graph. **[3]**

Edexcel June 2007 Unit Test 4

16 The Joint European Torus (JET) is a nuclear fusion experiment near Oxford in England. JET was the first experiment to produce a controlled nuclear fusion reaction.

a Describe the process of nuclear fusion. **[2]**

b Explain why it is difficult to maintain the conditions for nuclear fusion in a reactor. **[2]**

c The nuclei which fused were two isotopes of hydrogen. Why should the fusing of hydrogen nuclei release energy? **[2]**

Edexcel June 2007 Unit PSA5ii

[70 marks]

[Total 78 marks]

Experimental physics

Moving on from AS

ResultsPlus
Examiner tip

The *aim* of the experiment should guide everything you do. Keep the aim in mind all the time.

All A2 Physics students are required to carry out one piece of assessed practical work that is based on an application of physics. For your AS assessment you will have done a visit or a case study that led on to a practical that was fairly well defined. What is different at A2 is that you will be given a briefing that gives you a starting point and you will be asked to design your own experiment. All the skills you used at AS will be used again and mostly at a higher level.

Planning You will have much more freedom in your choice of apparatus than at AS, and your plan for the method will be entirely individual. Planning what you will actually do is the crucial new skill.

Implementation You are not expected to stick rigidly to your initial plan (see mark M4), if you have a better idea as you go on then by all means change your plan – but do write down what you are doing. You need to consider the effects of *uncertainty* much more thoroughly at A2, and that starts with your readings (see mark A2).

Analysis The relationship between the variables is more complex at A2. It might involve exponentials as well as squares or other powers. So your mathematical skills will be more important but these are something you can practise. You will need (for marks A12 and A14) to put uncertainties together to find their combined effect on your result – this is called '*compounding* uncertainties'. Also new at A2 is the need to reflect what else you might do to further your investigation and develop your ideas.

The briefing

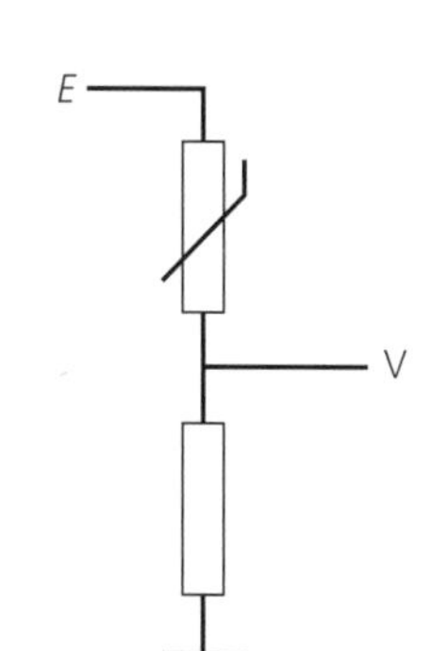

The briefing will contain details of the apparatus you are to use and a good indication of what you are investigating, but you are to plan how you will use the equipment and how you will draw a conclusion.

For example, in an investigation into a thermostat using a thermistor, the briefing gives the arrangement of the circuit elements shown here. You are also told the temperature range and given the mathematical model (or equation). Roughly speaking, the briefing will give you sufficient information to design a thermostat that works.

You will decide what apparatus you will use in order to vary and measure the temperature. Also you must decide on the electric circuitry you need.

Since it is a good idea to keep water and electricity apart, you will need to do a risk analysis to help ensure this.

As you can see, your task at A2 is more open-ended. Make sure you are clear about the aim of your practical work.

Planning

Use the assessment criteria to guide you, as with AS. Start with the last two points, P15 and P16, which are about producing a clear, structured plan. Think carefully about your subheadings for P15 and this will help with P16. Remember the aim of the experiment. Your grammar and spelling are important, too (P14), and since it is not possible to use a computer to write the plan you will need to pay attention to these as you write.

ResultsPlus
Build Better Answers

Good headings for your plan are: **Aim**, **Apparatus**, **Diagram and details**, **Method**, **Comments** (including possible subheadings *Safety*, *Using data*, *Uncertainties*).

Aim Start with the aim. Read the briefing carefully and write the aim down. Doing this at the start will help you to focus on it during your planning.

Apparatus As at AS, the description of apparatus is worth 4 marks (P4 to P7), but for P1 you will need to specify and explain your *own choice* of apparatus. Some will be suggested in the briefing, but what else you need and how you use it is up to you. Leave some space at the end of your list so that you can add to it without having to start again. Ensure that you list the measuring instruments you plan to use and explain your choices of two of these. The reason should be that the chosen ones give you the best precision, but you will need to explain that. Remember that *precision* is the smallest division on the measuring scale. So a multimeter on its 200 mA range will read to 0.1 mA since there are four figures in the display and its precision is 0.1 mA. Readings often fluctuate as you look at them, so the *uncertainty* is likely to be bigger than the precision but it cannot be smaller.

Diagram and details A good diagram (P3) will help you to collect your ideas together. You must be quite specific about details. If you need a beaker, specify that it will be a 250 cm^3 beaker, for example. Your diagram must have all the relevant dimensions shown, so it is worth spending a good few minutes drawing. Your diagram need not be a picture but it must show the essential apparatus and measurements. Any details not on your drawing can be written underneath. Make sure these are clear and include dimensions or values (mark P2).

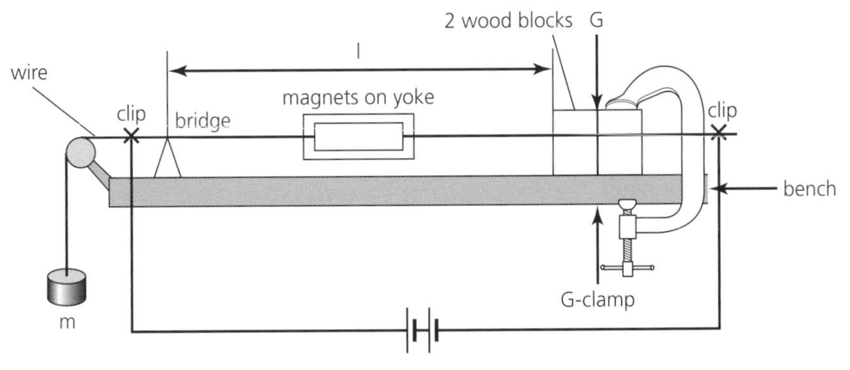

An example of a student's diagram

Method In planning your method, keep in mind the practical work you have done in the past.

You might need a good method for timing oscillations, for example. A counting mark at the centre is needed – an optical pin in a cork is good for this. To find a periodic time, record the time for ten oscillations and then divide by ten.

A 'difference method' is one where you measure a quantity before and after a change. This is what you do, for example, when finding the extension of a spring. The extension, or distance moved during the change, is found by taking the difference between your two readings.

You will need to think carefully about how you take your readings to make them accurate and precise. The method you adopt will be important, so think logically through the practical work. Start with your control variables and then plan how you will change your independent variable and observe the dependent one.

Comments Write your comments after your completed method. These should help you review your method and change it if necessary before you start work. Your comments should cover:

- P8 – your measuring techniques, such as the counting mark or difference method mentioned above. These could be shown on your diagram or noted in your method.
- P9 – how you will control *all* the other variables. There may be some variables that you cannot control, such as the room temperature, but you should mention it as a factor if you are doing a heating or cooling experiment.

Thinking Task

Experience shows that students who take care to repeat readings and have good methods for measuring always get more accurate results. Write down a good method to measure accurately:

a wavelength of a resonating wire,
b extension of a spring,
c temperature of a component in a beaker of water,
d time and current through a resistor as a capacitor decays.

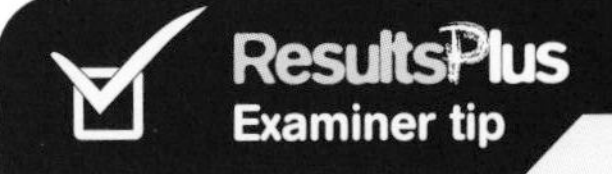

In order to make sure that a distance you are measuring is vertical or horizontal you can use a set square, or possibly two. Don't forget to show these on your diagram.

- P10 – why you will (or will not) be taking repeat readings. Remember that doing the experiment again is not repeating your readings but might still be a way to a more valid conclusion.
- P11 – all the *safety aspects*, including those that are minimal and saying why they are minimal. For example, a wire under tension requires the use of eye protection but a 12 V electricity supply is safe because the power supply is checked regularly by the school.
- P12 – what you will do with your data, for example your choice of graph to plot – see the Analysis section below.
- P13 – which of your measurements will have a *high uncertainty*, and point out what you will do to avoid systematic errors, such as checking the meters for zero error.

If you have a lot to say for one of the comments, you could use a separate subheading (such as *Safety*). The main thing is to remember to cover all the comments in a way that is clear.

Finally, read your plan again for spelling or grammar errors. You will now hand in your plan to be marked.

Quick Questions

Q1 Draw a diagram similar to the one on page 83 and add to it to show how you would measure the wavelength of a standing wave produced on the wire.

Q2 An experiment is set up to measure the force between two magnets as the distance between them is changed. Describe how you would measure the distance. A diagram will help.

Q3 State the instrument you would use and estimate the uncertainties in the following measurements. State any assumptions you make:

a the extension of a small spring as it stretches by about 30 mm,
b a voltage of 150 mV,
c the period of a pendulum whose period is about 1.5 s.

Write a risk assessment for an experiment you have done this year. It should be concise and say which risks pose a hazard.

Implementation

Remember that you must do all your planning *before* you do the practical work. You may modify what you do when you are actually doing the experiment, but your written plan will have been handed in beforehand and cannot then be altered. Your teacher may decide that you will carry out a different plan for the same experiment, but modified to help your teacher with the apparatus requirements.

Taking the measurements Read all your instruments carefully. For example:

- make sure your eye is level with the point on the scale you are looking at.
- will your time measurement be better if you use the lap-timer facility on the stop clock?

Recording your results Use a table with headings and units, as you did at AS.

- Make sure you quote your measurements – and subsequent calculations – to the appropriate number of significant figures.
- If you are taking repeats then record each reading and find a mean value. This will show the uncertainty.
- If you are not repeating measurements you could show an estimated uncertainty as a ± value at the bottom of a column.
- You should take enough readings to plot a good graph, and they should extend over as big a range as you can get.

Modifying your plan You are expected to refer to your written plan for mark M4. If you are going to change anything, you should note down exactly what you do – don't rely on memory. Then when you have finished, write under your table of results what you did differently. If your experiment is going well to plan, then record that no change was needed, and why.

If you are using a plan other than your own, you should expect to modify it and make decisions about the data handling.

Analysis

Graphs Much of your analysis will depend on drawing a graph. You must consider scales, axes and labels, plots and best-fit lines in the same way as you did for AS (marks A1 to A4). The biggest difference is that the variables in the A2 investigation will be linked in a complex way. The relationship is likely to involve a power or an exponential. As usual, a straight line is of most use so a little mathematics is needed to get this. A log-log graph is likely to be useful.

Worked Example

Suppose you are investigating a simple pendulum. The time period T and the length l are related by

$$T = 2\pi\sqrt{\frac{l}{g}}$$

so it follows that

$$T^2 = 4\pi^2\left(\frac{l}{g}\right) = \left(\frac{4\pi^2}{g}\right)l$$

Taking logs of both sides:

$$\log(T^2) = \log l + \log\left(\frac{4\pi^2}{g}\right)$$

but $\log(T^2) = 2\log T$

so $\log T = 0.5\log l + 0.5\log\left(\frac{4\pi^2}{g}\right)$

Compare this with

$$y = mx + c$$

This shows that a graph of log T against log l will give a straight line whose gradient is 0.5 and has an intercept that gives 0.5 log $\left(\frac{4\pi^2}{g}\right)$.

This works just as well for logarithms to base e, which we write as ln T and ln l.

ResultsPlus
Examiner tip

Remember that if the power is unknown then it comes from the gradient of the best-fit line of a log-log graph.

Worked Example

You are investigating the voltage V across a charged capacitor as it discharges through a resistor R.

Here we have $V = V_0 e^{-t/RC}$. This time we must take logarithms to base e. This gives:

$$\ln V = \frac{-t}{RC} + \ln V_0$$

Compare this with

$$y = mx + c$$

This shows that a graph of ln V against t gives a straight line with gradient $\frac{-1}{RC}$.

ResultsPlus
Watch out!

Show labels for the graph and the units in the results table in a form like this: log (T/s) or ln (l/mm).

If you always plot a graph using logarithms to base e, you will always be right. Be careful though, as you may only need a logarithm on the y-axis – compare the two Worked examples above.

You might have explained how you will plot your results as part of your plan, in which case you do not need to do it again. You will be credited with marks A5 and A6 there.

Uncertainties and errors Uncertainties occur because your readings fluctuate or are taken under slightly different conditions each time. *Errors* are more fundamental; they might occur for example because you ignored the effect of air resistance or the finite resistance of the voltmeter. For marks A10 and A12 you must show how the physics produces the equation you used and discuss whether the sources of error had a great effect.

For A13 you must calculate *percentage uncertainties*. Estimate the uncertainty in a reading and divide by the mean; it is enough to do this for one value only to get an estimate for that variable.

When you combine quantities such as voltage and current to find resistance, the uncertainties in the variables produce a combined uncertainty in the final answer. You must *compound* the uncertainties (A14).

- If the two quantities are multiplied together or divided, you add the percentage uncertainties. For example, if the uncertainty in the voltage readings is 2% and in the current readings is 3% then the uncertainty in the resistance is (3 + 2)% = 5%.
- When a quantity is squared the percentage uncertainty is doubled; if the quantity is cubed you triple the percentage uncertainty. In other words, you multiply the percentage uncertainty by the power. So if the uncertainty in T is 3% the uncertainty in T^2 is 6%.

This bar represents the uncertainity in the variable plotted on the y-axis

data point

This bar represents the uncertainity in the variable plotted on the x-axis

How to draw error bars

Error bars Error bars are used on graphs to show the measurement uncertainty in your data points. To plot them you add vertical and horizontal lines to each data point which are the length of the uncertainty in each variable. You can use the same value for all the points for each variable, but remember that the uncertainties in the two variables you have plotted may not be the same. You can obtain an uncertainty in your value for the gradient of a best-fit line by drawing a 'worst-fit line' that just passes through *all* the error boxes and measuring its gradient. The difference between the two gradient values is the uncertainty in your gradient.

For example, in an experiment to find the Young modulus E, we might plot load F against extension x to give a gradient value of m. From the equation $E = \frac{Fl}{xA}$ we have $F = \left(\frac{EA}{l}\right)x$, and so $\frac{EA}{l}$ is the gradient m of the line F against x, and $E = \frac{ml}{A}$. The percentage uncertainty in E is then the sum of the percentage uncertainties in m and l and A.

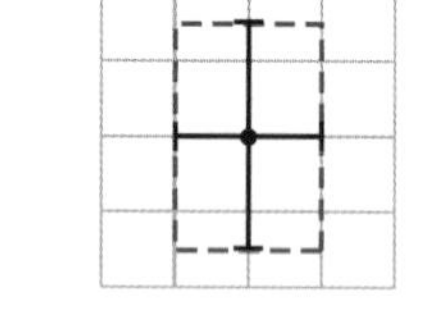

Use an error box to help you plot your worst-fit line

Remember that you are trying to get an estimate for the total uncertainty based on your readings. This allows you to gauge how accurate your result is.

Quick Questions

Q4 The time T for an oscillation varies with the length d as $T = kd^n$. You take readings of T and d. What graph do you plot to find the value for n?

Q5 You measure the diameter of a cylinder as 5.4 ± 0.2 cm and the height as 24.0 ± 0.2 cm. Calculate the volume of the cylinder and the uncertainty in the volume.

Q6 What is the difference between a precise result and an accurate one?

Conclusion

State your final value and compare it with the value given, if one has been suggested. Review your experiment (for A17) by discussing how well you have achieved the aim and, for A16, state your conclusion clearly. This discussion should mention the uncertainties; if these are large then your result is not very reliable. Bring in the effect of any errors, too. For A11 you should consider the precision of your result and also its accuracy – make sure you can use these terms correctly. You should, for A15, consider how the investigation might be improved to reduce the effect of any errors or to reduce your uncertainties. Simply saying 'more time...' is not likely to be enough – a suggestion for changing the way you do the practical work is expected. You can suggest any apparatus you want, such as a data logger, but you must say *why* this will make the result more reliable and/or accurate.

Finally, how might you develop the investigation, for mark A18? You can broaden the aim or change the focus but remember to think about the physics behind your investigation and how you can find out more about that.

Answers

Answers to quick questions: Unit 4

Momentum and force

1 1660 N s; assumption: no work done against resistive forces
2 -1.6×10^4 N s
3 15 N

Thinking Task
Crumple zones increase the collision time, hence the collision force is reduced for the same momentum change.

Momentum and collisions

1 9.0 m s^{-1}
2 **a** 6.2 m s^{-1} **b** elastic, because K.E. is conserved
3 100 m s^{-1}

Thinking Task
21 m s^{-1}, 49° 51% lost lost K.E. mainly becomes P.E. of deformation and thermal energy

Circular motion 1

1 0.95 rev min^{-1}
2 7.27×10^{-5} rad s^{-1} 0.22 m s^{-2}
3 1.05 m

Thinking Task
a 6.7×10^{-4} rad s^{-1} **b** 4.3 m s^{-2}

Circular motion 2

1 2500 N
2 **a** Car's direction is changing, and so it is accelerating
b Towards the centre of the circular path
c 13.8 m s^{-1}
3 910 N

Thinking Task
a 130 N **b** 10°

Electrostatics

1 8.2×10^{-8} N
2 6.4×10^{-13} N
3 6.38×10^5 C, negative

Thinking Task
a

Distance from point charge/cm	Force on sphere/μN
1.00	90.0
2.00	**22.5**
1.73	30.0

b 2.5×10^{-9} C

Capacitors

1 2.5×10^{-17} C
2 14000 V
3 6×10^{-11} F

Thinking Task
37 J As charge flows in the discharging process the p.d. across the capacitor falls, making the capacitor only suitable to provide a burst of energy.

Exponential changes

1 2.7×10^{-4} F
2 $\Omega = \text{V A}^{-1}$, $\text{F} = \text{C V}^{-1}$,
so RC has units $\Omega \times \text{F} = \text{V A}^{-1} \times \text{C V}^{-1} = \text{C A}^{-1}$.
C = A s, hence RC has units A s A^{-1} = s.
3 1.6 s

Thinking Task
95 μF

Magnetic fields

1 Field lines indicate the direction of the force on a north pole; the force has one direction at any point in the field.
2 12 N
3 With magnetic field into the paper and electrons initially travelling bottom to top, the beam is initially deflected to the right.

Thinking Task
a 7.5×10^{10} C kg^{-1}
b Electron is travelling near to the speed of light and therefore has an increased mass (relativistic effects).

Generating electricity

1 2×10^{-5} V
2 1.5×10^{-3} V
3 There is only air resistance opposing the steel bar as it falls through the tube.
As the bar magnet drags flux down the tube it induces an emf near to each pole (Faraday's law).
The induced emfs cause currents to circulate in the tube near to each pole.
According to Lenz's law these currents must be so as to oppose the flux changes.
The currents cause a magnetic force to act on the magnet in the opposite direction to its motion.
Hence the magnet accelerates at a lower rate, as the total force opposing its fall is greater.

Thinking Task
Current in the coil in the charger creates a magnetic field that links with the coil in the base.
An alternating voltage applied to the base coil will generate an alternating magnetic field.
The flux linkage change for the base coil induces an emf according to Faraday's law.

The nuclear atom

1 Each force away from nucleus along a line joining alpha particle to nucleus, largest force for closest approach
2 **a** $^{236}_{92}\text{U} \rightarrow {}^{144}_{56}\text{Ba} + {}^{89}_{36}\text{Kr} + 3 \times {}^{1}_{0}\text{n}$
b $^{144}_{56}\text{Ba} \rightarrow {}^{144}_{57}\text{La} + {}^{0}_{-1}\beta^- + {}^{0}_{0}\nu_e$
3 7.3×10^{-11} m

Thinking Task
$v = 3.5 \times 10^7$ m s^{-1} $\lambda = 2.1 \times 10^{-11}$ m unsuitable because wavelength is much too big

High energy collisions

1 To gain energy each time they cross the gap between the drift tubes, the time taken for the particles to

move along each tube must match the time period of the alternating supply.

As the particles speed up, the time taken to cover a fixed distance decreases.

Hence the particles would arrive at the gap before the next tube had the appropriate polarity, and so they would lose energy as they crossed the gap.

2 21.9 GeV

Thinking Task

6.5 MeV

Particle accelerators

1 5.8×10^7 m s^{-1}

2 $r = \frac{p}{BQ}$ and $p = mv$ so $E_k = \frac{1}{2}mv^2 = \frac{1}{2}p^2/m$; $E_k = \frac{B^2Q^2r^2}{2m}$

3 0.038 m

Particle theory

1 ${}^1_1p \rightarrow {}^1_0n + {}^0_1e^+ + {}^0_0\nu_e$ an up quark changes to a down quark

2 a A baryon is a hadron consisting of 3 quarks.

b $\lambda^0 \rightarrow n + \pi^0$

41 MeV

Thinking Task

a Bar over the b indicates that it is an anti-bottom quark

$u\bar{b}$ as charge must be 1; $\bar{b}$ has charge $\frac{1}{3}$, u has charge $\frac{2}{3}$

b 4.3 GeV

Answers to quick questions: Unit 5

Internal energy

1 a $\frac{T}{K} = \frac{t}{°C} + 273 = 1085 + 273 = 1358$

$\Rightarrow T = 1358\,K$

b $\frac{T}{K} = \frac{t}{°C} + 273 \Rightarrow \frac{t}{°C} = \frac{T}{K} - 273 = 2 - 273$

$= -271 \Rightarrow t = -271°C$

2 A and B must be at the same temperature.

Thinking Task

a For a fixed temperature difference between the radiator and the room, a certain mass of water carries more energy than the same mass of any other liquid. Put another way: water can carry a lot of energy around without its temperature changing too much.

b Estimates: a cup of tea needs 200 g (0.2 kg) of water. Initial temperature of water = 20°C.

$\Delta E = mc\Delta\theta \Rightarrow P \times t = mc\Delta\theta \Rightarrow$

$t = \frac{mc\Delta\theta}{P} = \frac{0.2 \times 4200 \times (100-20)\,J}{3000\,W} = 22.4\,s$

Gas laws and kinetic theory

1 $pV = NkT \Rightarrow k = \frac{pV}{NT}$

This fraction has units as follows:

$\frac{Pa\,(N\,m^{-2}) \times m^3}{(\text{no unit}) \times K} \Rightarrow \frac{N\,m}{K} \Rightarrow J\,K^{-1}$

2 $\frac{1}{2}m\langle c^2\rangle = \frac{3}{2}kT$

Consider square-rooting both sides: the r.m.s.speed is proportional to the square root of T. So if T increases by a factor of 2 (300 K to 600 K), then the r.m.s. speed increases by a factor of $\sqrt{2}$, that is 1.4.

3 $\frac{p_2V_2}{T_2} = \frac{p_1V_1}{T_1}$

T is constant here, so

$V_2 = \frac{p_1V_1}{p_2} = \frac{3.50 \times 10^5\,Pa \times 0.02\,m^3}{1.00 \times 10^5\,Pa} = 0.07\,m^3$

Thinking Task

very well: 1, 3; pretty well: 4, 5; not at all well: 2

Nuclear radiation

1 Use $A = -\lambda N$ and $\lambda = \frac{\ln 2}{t_{1/2}} \Rightarrow A = N \times \frac{\ln 2}{t_{1/2}}$

for X: $A = 2 \times 10^{20} \times \frac{\ln 2}{4 \times 10^6 \times 365 \times 24\text{ hours}}$

$= 4.0 \times 10^9\,\text{hour}^{-1}$

for Y: $A = 5 \times 10^9 \times \frac{\ln 2}{6\text{ hours}} = 5.8 \times 10^8\,\text{hour}^{-1}$

$\Rightarrow$ X is more active

2 a $\lambda = \frac{\ln 2}{t_{1/2}} = \frac{\ln 2}{7.1\,s} = 0.0976\,s^{-1}$

b $N = N_0e^{-\lambda t} \Rightarrow t = \ln\left(\frac{N}{N_0}\right) \div \lambda = \ln\left(\frac{1.0\,mg}{5.0\,mg}\right)$

$\div 0.0976\,s^{-1} = 16.5\,s$

Thinking Task

a To create an ion the radiation has to give the ionised atom some energy. So if it's good at ionising, as an α is, it will lose its energy quickly, and not penetrate much matter. Conversely, a γ hardly ionises at all, so penetrates a long distance.

b One throw of the dice represents a period of time. Dice and unstable nuclei both behave randomly. In one throw a certain fraction of dice – expected to be one sixth – 'decays'. In a fixed period of time a certain fraction of the nuclei decays.

Binding energy, fission and fusion

1 Mass of nucleons = $26 \times m_p + 28 \times m_n$

$= 26 \times 1.00728\,u + 28 \times 1.00867\,u = 54.43204\,u$

$\Rightarrow \Delta m = 54.43204\,u - 53.93962\,u = 0.49242\,u$

$= 0.49242 \times 1.66 \times 10^{-27}\,kg = 8.17 \times 10^{-28}\,kg$

$\Rightarrow \Delta E = c^2\Delta m = (3.00 \times 10^8\,m\,s^{-1})^2 \times 8.17 \times 10^{-28}\,kg = 7.36 \times 10^{-11}\,J$

$= \frac{7.36 \times 10^{-10}\,J}{1.6 \times 10^{-13}\,J\,MeV^{-1}} = 460\,MeV$

$\Rightarrow$ binding energy per nucleon for

${}^{54}_{26}Fe = \frac{460\,MeV}{54} = 8.51\,MeV$

2 a A **b** B **c** C **d** C **e** D **f** C **g** A

Thinking Task

a $4{}^1_1H \rightarrow 4{}^4_2He + 2{}^0_1e + 2\bar{\nu}_e$

b Initial mass = $4 \times m_p = 4 \times 1.00728\,u = 4.02912\,u$

Final mass = $4.00260\,u + 2 \times 0.00055\,u = 4.00370\,u$

$\Rightarrow \Delta m$ = 4.02912 u − 4.00370 u = 0.02542 u
= $0.02542 \times 1.66 \times 10^{-27}$ kg = 4.22×10^{-29} kg
$\Rightarrow \Delta E = c^2\Delta m = (3.00 \times 10^8\,\mathrm{m\,s^{-1}})^2 \times 4.22 \times 10^{-29}$ kg
= 3.80×10^{-12} J
This is the energy from one fusion event, which uses four hydrogen nuclei.
$\Rightarrow$ There are $6.02 \times 10^{26} \div 4$ events
= 1.505×10^{26} events
$\Rightarrow$ energy released = $3.80 \times 10^{-12}\,\mathrm{J} \times 1.505 \times 10^{26}$
= 5.72×10^{14} J

Simple harmonic motion

1 a $\omega = \frac{2\pi}{T} = \frac{2\pi}{3.0\,\mathrm{s}} = 2.1\,\mathrm{rad\,s^{-1}}$

b maximum speed = $A\omega$ = $0.2\,\mathrm{m} \times 2.1\,\mathrm{rad\,s^{-1}}$
= $0.42\,\mathrm{m\,s^{-1}}$

c maximum acceleration = $A\omega^2$ = $0.2\,\mathrm{m} \times (2.1\,\mathrm{rad\,s^{-1}})^2$
= $0.88\,\mathrm{m\,s^{-2}}$

2 First calculate ωt at 2.0 s: $\omega t = 7.4\,\mathrm{rad\,s^{-1}} \times 2.0\,\mathrm{s}$
= 14.8 rad
Then $x = A\cos\omega t = 2.5 \times 10^{-3}\,\mathrm{m} \times \cos(14.8\,\mathrm{rad})$
= -1.5×10^{-3} m
$v = -A\omega \sin\omega t = -2.5 \times 10^{-3}\,\mathrm{m} \times 7.4\,\mathrm{rad\,s^{-1}}$
$\times \sin(14.8\,\mathrm{rad}) = -1.5 \times 10^{-2}\,\mathrm{m\,s^{-1}}$
$a = -A\omega^2\cos\omega t = -2.5 \times 10^{-3}\,\mathrm{m} \times (7.4\,\mathrm{rad\,s^{-1}})^2$
$\times \cos(14.8\,\mathrm{rad}) = -8.4 \times 10^{-2}\,\mathrm{m\,s^{-2}}$

Thinking Task

a No. While the person is in the air they experience a constant force downwards due to gravity – not a force proportional to displacement. When in contact with the ground they experience a large upward force for a short time.

b Probably yes. Provided the elastic obeys Hooke's law and never goes slack, the force will be proportional to the displacement and in the opposite direct, as needed for SHM.

c No. While they are in contact with the trampoline the force might satisfy the conditions for SHM, but in the air the force is constant as in **a**.

Energy and damping in SHM

Thinking Task

$E_{tot} = \frac{1}{2}kA^2$ so

a A doubles $\Rightarrow E_{tot}$ is multiplied by 4
b m doubles $\Rightarrow E_{tot}$ is unaffected
c k doubles $\Rightarrow E_{tot}$ doubles.

1 a maximum speed is $A\omega$, $\omega = \frac{2\pi}{T}$

$\Rightarrow$ maximum $E_k = \frac{1}{2} \times m \times (\text{maximum speed})^2$
$= \frac{1}{2}mA^2\omega^2$
$= \frac{1}{2} \times 0.2\,\mathrm{kg} \times (0.4\,\mathrm{m})^2 \times \left(\frac{2\pi}{3.0\,\mathrm{s}}\right)^2 = 0.070\,\mathrm{J}$

b Use the fact that E_{tot} = maximum E_k, since then E_p = 0
$\Rightarrow \frac{1}{2}kA^2 = 0.070\,\mathrm{J} \Rightarrow k = \frac{2 \times 0.070\,\mathrm{J}}{A^2}$
$= \frac{2 \times 0.070\,\mathrm{J}}{(0.4\,\mathrm{m})^2} = 0.88\,\mathrm{N\,m^{-1}}$

c When E_k and E_p are equal, then each is equal to half of E_{tot}.
Then $E_p = 0.035\,\mathrm{J} = \frac{1}{2}kx^2 \Rightarrow x^2 = \frac{2 \times 0.035\,\mathrm{J}}{k}$
$= \frac{2 \times 0.035\,\mathrm{J}}{0.88\,\mathrm{N\,m^{-1}}} = 0.0798\,\mathrm{m^2}$
$\Rightarrow x = 0.28$ m

2 E_{tot} = maximum $E_k = \frac{1}{2}mA^2\omega^2 \Rightarrow A^2 = \frac{2 \times E_{tot}}{m\omega^2}$
$= \frac{2 \times 5.7 \times 10^{-20}\,\mathrm{J}}{1.7 \times 10^{-27}\,\mathrm{kg} \times (2 \times \pi \times 8.7 \times 10^{13}\,\mathrm{s^{-1}})^2} \Rightarrow$
$A = 1.5 \times 10^{-11}$ m

Forced oscillations and resonance

1 a A particular speed means a particular driving frequency. Only one speed will drive the mirror at its natural oscillation frequency, so that at this driving speed it will resonate, giving it noticeable amplitude of vibration.

b The Blu-tack adds mass, so the natural frequency is reduced. Perhaps it becomes lower than the car ever revs at.

2 a Light damping gives a very sharp resonance graph, which enables the radio to select one frequency very precisely.

b The electrical resistance in the circuit absorbs energy. Make the resistance as low as possible.

3 a If $f >> f_o$ then we are on part of the response graph where the amplitude of the driven oscillations is very small.

b When machinery is first switched on its rotation rate rises from zero. So the driving frequency f rises, at some stage passing through the value of f_o.

Thinking Task

The mass can also swing as a pendulum. And each length of string can have standing waves set up along it. Each of these modes of oscillation has its own frequency.

Gravity and orbits

1 a $r^3 = \frac{GMT^2}{4\pi^2} = \frac{6.67 \times 10^{-11}\,\mathrm{N\,m^2\,kg^{-2}} \times 5.99 \times 10^{24}\,\mathrm{kg} \times (27.3 \times 24 \times 60 \times 60\,\mathrm{s})^2}{4\pi^2}$

$= 5.63 \times 10^{25}\,\mathrm{m^3} \Rightarrow r = 3.83 \times 10^8$ m

b $3.83 \times 10^8\,\mathrm{m} \div 6.38 \times 10^6\,\mathrm{m}$ = 60 Earth's radii.

c $g = \frac{Gm_1}{r^2} = \frac{6.67 \times 10^{-11}\,\mathrm{N\,m^2\,kg^{-2}} \times 5.99 \times 10^{24}\,\mathrm{kg}}{(3.83 \times 10^8\,\mathrm{m})^2}$
$= 2.72 \times 10^{-3}\,\mathrm{N\,kg^{-1}}$

d $g = \frac{F}{m}$ but also $\frac{F}{m} = a$, so g is also the acceleration of the Moon.

2 a B **b** A **c** C **d** C **e** A **f** B

Thinking Task

$g = \frac{GM}{r^2}$ M = volume × density. Volume is multiplied by four-cubed; density is halved; so M is 32 times larger But r-squared is four-squared times larger; so the overall effect on g is to increase it by $\frac{32}{16}$ times. Answer: A.

Stars

1 a $T^4 = \frac{L}{4\pi r^2 \sigma}$

$= \frac{1.9 \times 10^{25}\,W}{4\pi \times (4.2 \times 10^8\,m)^2 \times 5.67 \times 10^8\,W\,m^2\,K^{-4}}$

$= 1.51 \times 10^{14}\,K^4 \Rightarrow T = 3500\,K$

b $\lambda_{max} T$ = constant (2.898×10^{-3} m K) $\Rightarrow$

$\lambda_{max} = \frac{2.898 \times 10^{-3}\,m\,K}{3500\,K} = 8.3 \times 10^{-7}\,m$

2 a $F = \frac{L}{4\pi d^2} = \frac{1.9 \times 10^{25}\,W}{4\pi \times (11.6 \times 9.5 \times 10^{15}\,m)^2}$

$= 1.2 \times 10^{-10}\,W\,m^{-2}$

b From another galaxy the distance away of the star and the Sun are near enough equal. So the lower luminosity of the star will mean it is less bright. Its λ_{max} is significantly greater than that of the Sun, so it will appear more red.

3 $L \approx 3 \times 10^5 L_{Sun}$

The universe

1 $\Delta\lambda$ = 512 nm – 434 nm = 78 nm → $z = \frac{\Delta\lambda}{\lambda}$

$= \frac{78\,nm}{434\,nm} = 0.180$

$v = cz = 0.180 \times 3 \times 10^8\,m\,s^{-1} = 5.39 \times 10^7\,m\,s^{-1}$

$= 5.39 \times 10^4\,km\,s^{-1}$

$d = \frac{v}{H_o} = \frac{5.39 \times 10^4\,km\,s^{-1}}{71\,km\,s^{-1}\,Mpc^{-1}} = 759\,Mpc$

2 a $H_o = 71\,m\,s^{-1}\,Mpc^{-1} = \frac{71\,km\,s^{-1}}{1\,Mpc}$

$= \frac{71 \times 10^3\,m\,s^{-1}}{1 \times 10^6 \times 3.09 \times 10^{16}\,m} = 2.30 \times 10^{-18}\,s^{-1}$

b Age $= \frac{1}{H_o} = \frac{1}{2.30 \times 10^{-18}\,s^{-1}} = 4.34 \times 10^{17}\,s$

$= 1.4 \times 10^{10}$ years

c This is an estimate because it assumes uniform expansion. We do not know whether the expansion has been slowing due to gravity forces reducing with distance; recent evidence actually indicates a speeding up of the expansion.

Thinking Task

a A closed universe is one in which there is sufficient matter eventually to stop the expansion and make the universe contract.

b When the expansion stops, all galaxies will have zero redshift. During the contraction, blueshifts will be observed, as galaxies are approaching. The amount of blueshift will be proportional to their distance away as before. As they approach faster and faster the amount of blueshift will increase.

Answers to quick questions: Unit 6

Thinking Task

a Ensure resonance sharp; hold rule close to wire; look with eye perpendicular to wire at point of reading (no parallax); read distance between as many nodes as possible; calculate mean distance between nodes; wavelength is double this.

b Metre rule vertical using set square; rule close to spring; eye perpendicular to rule at point of reading (no parallax); read position of bottom of spring before and after loading; extension is difference in two positions (difference method).

c Thermometer close to component; stir liquid during heating; allow time for component to heat through.

d After trial run draw table and insert time values; place clock close to meter; count down to start 3, 2, 1, 0, 1 etc. and start clock on zero as switch is closed; tap with pencil as seconds reach desired value in table; as reading time approaches look from clock to meter still tapping.

1 You should show the measurement including as many nodes as possible. If you measure the distance to include 5 nodes then that is two wavelengths.

2 Hold a metre rule vertically (with a set square) and look across the surface of the magnet to determine its position and repeat for the other magnet. Calculate the distance between the positions of the two surfaces.

3 a Probably 1 mm uncertainty at each position so 2 mm in the extension. Assume the rule is vertical and the eye line is perpendicular to the scale.

b Check for zero error. On 200 mV scale, uncertainty is probably 0.5 mV, reading probably fluctuates. Assuming temperature of the components is not changing and resistance of meter high enough not to affect reading.

c Measure 10 periods and divide uncertainty in reading by 10 giving probably 0.05 s or better. Assume a counting marker was at the centre of the oscillation, and period independent of amplitude, i.e. amplitude small.

Thinking Task

Your risk assessment should detail the hazard (i.e. what might damage you or the apparatus). It should determine how likely this is to occur.

If there is no risk you should say why there is no risk (e.g. 12 V is too low to cause a fire or electrocution but you must check there are no (accidental) short circuits).

4 You plot ln T vs ln d and the slope of the straight line will be the value for n.

5 $\frac{\pi}{4} \times (5.4)^2 \times 24.0 = 550\,cm^3$ – to 2 or 3 significant figures (SF).

Uncertainty $= 2 \times \frac{\delta d}{d} + \frac{\delta h}{h} = 2 \times \left(\frac{0.2}{5.4}\right) + \left(\frac{0.2}{24}\right)$

$= 2 \times 0.037 + 0.0083$

$= 0.082$ or 8%

So uncertainty in volume = $550 \times 0.082 = 45\,cm^3$.

Volume = $5.50 \times 10^{-4}\,m^3 \pm = 0.45 \times 10^{-4}\,m^3$

6 A precise result is one that correctly quotes a large number of significant figures and a small uncertainty. An accurate result is one that is close to the true value.

Answers to practice exam questions: Unit 4

Section 1 Further mechanics

1 D **2** D **3** D **4** C

5 a Momentum is conserved [1], so he must recoil with a momentum equal in magnitude to the forward momentum of the ball [1].

b $v = 2.5 \times \frac{12.5}{60}$ [1], 0.52 m s^{-1} [1]

6 Momentum before: 0 [1]
Momentum of gun after: 350g × v [1]
Momentum of pellet after: 2.5g × 12 m s^{-1} [1]
$v = 2.5\text{g} \times \frac{12\text{ m s}^{-1}}{350} = 0.086\text{ m s}^{-1}$ [1]

7 Momentum of ball before:
6.25 kg × 5.5 m s^{-1} = 34.4 N s [1]
KE before: 0.5 × 6.25 kg × (5.5 m s^{-1})^2 = 94.5 J [1]
KE conserved so $3.125 v_{ball}^2 + 0.6 v_{skittle}^2 = 94.5$ [1]
momentum convserved so
$34.4 = 6.25\text{kg} \times v_{ball} + 1.2\text{kg} \times v_{skittle}$ [1]
rearranging gives v_{ball} in terms of $v_{skittle}$; substitute into KE equation to give $1.43 v_{skittle}^2 = 13.2 v_{skittle}$ [1]
and so $v_{skittle} = \frac{13.2}{1.43} = 9.2\text{ m s}^{-1}$ [1]

8 a The diagram is to help with getting directions correct in the calculation in **b**

b Momentum before: $6.7 \times 10^{-27} \times 1.2 \times 10^{7}$ $= 8.04 \times 10^{-20}$ Ns [1]
Momentum after: $-6.7 \times 10^{-27} \times 9.0 \times 10^{6}$ $+ 4.64 \times 10^{-26}\, v_N$ [1]
Rearranging, $v_N = 1.40 \times \frac{10^{-19}}{4.64} \times 10^{-26}$
$= 3.0 \times 10^{6}$ m s^{-1} [1]

c KE before: $0.5 \times 6.7 \times 10^{-27} \times (1.2 \times 10^{7})^2$ $= 4.8 \times 10^{-13}$ J [1]
KE after: $0.5 \times 6.7 \times 10^{-27} \times (9 \times 10^{6})^2 + 0.5 \times 4.64 \times 10^{-26} \times (3 \times 10^{6})^2$
These are the same, so the collision is elastic [1]

9 a Momentum before: 65 × 25 [1] = 1630 N s [1]

b $F = \frac{\text{change in momentum}}{\text{time taken}} = \frac{1630}{(75 \times 10^{-3})}$ [1]
$= 2.2 \times 10^{4}$ N [1]

c Because the seatbelt is not the only mechanism slowing the driver down.

10 a $v = \omega r = 900 \times 2\pi \times \frac{0.25}{60}$ [1] = 23.6 m s^{-1} [1]

b use $a = \frac{v^2}{r}$ to give 2220 m s^{-1} [1]; inwards to center of drum [1]

c Push on shirt from inside of drum

d No force on water from drum where there are holes in the drum, so water carries on moving in a straight line (Newton's 1st law) and escapes through holes in drum.

11 a Centripetal force needed to keep rider moving in a circlar path. As chair swings out (actually maintaining a straight line path according to Newton's 1st law) a component of the tension in the chains acts towards the centre of the circle.

b resolve forces vertically; $T\cos\theta = mg$ so $\cos\theta$ $= 45 \times \frac{9.81}{550}$ [1], $\theta = 36.7°$ to the vertical [1]

c T is found by resolving forces horizontaly, $T = m\frac{v^2}{r}$ [1], so in the expression for $\cos\theta$ the mass of the rider cancels out [1]

d 8.5 m s^{-1}

e The angle increases as the rides speeds up, since a greater centripetal force is required at higher speeds, and so the horizontal component of the tension must increase.

12 a $\omega = \frac{2\pi}{4.5} = 1.4$ rad s^{-1} [1]

b $a = \omega^2 r = 1.4$ m s^{-2} [1]

c $F = ma$ for each girl [1] $F = 65 \times 1.4 = 9.1$ N [1]

d They will move at a speed of 1.0 m s^{-1} [1 for calculation of $v = \omega r$] travelling in a straight line at a tangent to the circle [1] at the instant that they let go.

Section 2 Electric and magnetic fields

1 A **2** B **3** D **4** C

5 a $E_q = mg$ [1] and $E = 6050$ V m^{-1} [1], so $Q = mg/E$ [1] 9.6×10^{-19} C [1]

b 6 [1]

6 a $\frac{1}{2}mv^2 = QV$ [1] $v = \sqrt{2} \times 1.6 \times 10^{-19}$ C $\times 230 \frac{\text{V}}{9.1} \times 10^{-31}$ kg [1] $= 1.1 \times 10^{7}$ m s^{-1} [1]

b i vertical parallel equally spaced lines away from positive plate [1]

ii beam is deflected upwards in parabolic path [1]

c $E = \frac{V}{d}$ [1] 3000 V m^{-1} [1]

d perpendicularly into plane of paper [1], applying Fleming's left hand rule (downward force is needed, current is right to left) [1]

e For no deflection the magnetic and electric forces balance so $Bev = Ee$ [1], $B = \frac{E}{v} = \frac{V}{vd}$ [1]
2.7×10^{-4} T [1]

7 a i 3.3×10^{-5} C, 9.9×10^{-5} C [2]

ii 2.5×10^{-4} J, 1.5×10^{-3} J [2]

b capacitors are in parallel [1]

c [use total charge remains constant [1], and ratio of charges equals ratio of capacitances [1]], 5.3×10^{-5} C, 7.9×10^{-5} C [1], 24 V [1]

d 1.6×10^{-3} J, work is done in redistributing charge [3]

8 use of $V = V_0 e^{-t/RC}$ [1]; $t = 60$ s [1]; correct answer (14 kΩ) [1]

9 a i Magnet takes only a fraction of a second to pass through the coil [1] so data must be collected over very short time interval [1]

ii South pole [1]

b Negative voltage is when magnet is leaving coil; travelling faster at this stage [1], so rate of change of flux linkage is greater (hence induced emf is larger) [1]

c Area between graph and time axis represents the flux linked with the coil [1]. Total flux linked after magnet has left coil must be zero, so positive and negative areas should cancel [1].

10 As disc rotates it cuts magnetic flux [1], this induces an emf [1] (Faraday's law) to make a current flow [1] to oppose [1] the flux cutting (Lenz's law). Hence there is a resistive force on the disc, slowing the disc down. [1]

11 a Soft iron rod becomes magnetised [1], channelling magnetic flux so that (almost) all flux produced by primary coil links with the secondary coil [1].

b i A steady current produces a constant magnetic field, so the rate of change of flux linkage is zero [1], hence there is no induced emf [1] (Faraday's law).

ii The primary coil has a very small resistance, so the current will be quite large, causing quite a large rate of dissipation of energy [1]. When a.c. is applied an emf will be induced in the primary to oppose the changes in flux linkage (Lenz's law), hence the net p.d. in the primary will be (almost) zero [1]. Rate of dissipation of energy is greatly reduced [1].

Section 3 Particle physics

1 D **2** B **3** B **4** C

5 a $^{232}_{90}\text{Th} \rightarrow\ ^{228}_{88}\text{Ra} +\ ^{4}_{2}\alpha$ [1 for alpha correct, 1 for radium correct]

b $^{208}_{81}\text{Tl} \rightarrow\ ^{208}_{82}\text{Pb} +\ ^{0}_{-1}\beta^- +\ ^{0}_{0}\bar{\nu}_e$ [1 for beta correct, 1 for thallium, 1 for antineutrino]

c udd → uud (d → u) [1], lepton [1]

d Only α decay changes nucleon number [1], therefore 6 α-particles [1], 4 β-particles to balance protons [1]

6 a Magnetic force on electrons acts perpendicularly to plane containing B and v [1]. Force always at right angles to velocity, producing a centripetal acceleration [1].
Magnetic field must be perpendicular into plane of paper [1].

b Use of $W = QV$ [1]; 6.7×10^{-15} J [1]

c Use of $mv = BQr$ [1] Use of $f = \dfrac{BQ}{2\pi m}$ [1] 2.5×10^7 Hz [1]

d 1.0×10^{-11} J [1] Use of $r = \dfrac{p}{BQ}$ [1] 0.69 m [1]

e As the protons approach the speed of light their mass will increase. [1] The time taken for the protons to make half a revolution depends upon the mass, so the protons will not stay synchronised [1] with the alternating voltage applied between the dees.

7 a The π^0 particle is uncharged and so doesn't produce ionisation [1]

b 2 tracks to the right [1] (apply Fleming's left hand rule) [1]

c Least curved track (top track) has higher energy [1], since $r = \dfrac{p}{BQ}$ [1]

d Both are composed of quarks [1]. Protons consist of 3 quarks (baryon), pions consist of a quark-antiquark pair (meson) [1].

e 2.2 GeV [1]

8 a Use of $\lambda = \dfrac{h}{mv}$ [1] $2.9 \times 10^7\ \text{m s}^{-1}$ [1]

b Use of energy conservation [1] Substitute values for e and m [1] 2.4×10^3 V [1]

c The spacing of atoms in graphite is ~ 10^{-11} m [1]; for an acceptable diffraction pattern the wavelength needs to match to the size of the diffracting aperture [1].

9 a Plum pudding model: uniform sphere of positive charge dotted with electrons [1].
Nuclear model: Nucleus with all the positive charge and most of the mass of the atom [1], electrons circle about nucleus [1], nucleus is a tiny fraction of the atomic size [1].

b i Most αs passed through with minimal deflection [1], some αs were scattered by large angles [1], a tiny fraction of αs bounced back from the foil [1].

ii The αs that were scattered through large angles must have experienced a very strong repulsive force [1]; Rutherford showed that this could only be if the positive charge of the atom were concentrated in a tiny fraction of the atom's volume [1].

Answers to practice exam questions: Unit 5

Section 4 Thermal energy

1 C

2 B

3 T is proportional to average kinetic energy. So the warmer distribution of molecules (sea level) is the one with its peak to the right [1] – where the average kinetic energy is greater [1].

4 a $\Delta E = mc\Delta\theta$ [1] $= 8.70 \times 10^{-5}\ \text{kg} \times 385\ \text{J kg}^{-1}\,°\text{C}^{-1} \times (1080 - 20)\,°\text{C} = 35.5\ \text{J}$ [1]

b $\Delta E = P \times t \Rightarrow t = \dfrac{\Delta E}{P} = \dfrac{35.5\ \text{J}}{2.2\ \text{W}} = 16.1\ \text{s}$ [1]

c Possible discussions: As fuse heats up, resistance rises, so rate of heating increases [1] (if current constant), **so time reduces** [1]; OR ... As fuse heats up, rate of heat loss from surface increases [1], **so time increased** [1].

5 a Absolute zero is the temperature at which the kinetic energy of the molecules in the material is zero OR ... the temperature at which an ideal gas would exert zero pressure OR ... an ideal gas would occupy zero volume [1].

b i $pV = NkT$ [1]

$\Rightarrow N = \dfrac{pV}{kT}$ [1]

$= \dfrac{(1.1 \times 10^5 + 1.0 \times 10^5)\ \text{Pa} \times 5.8 \times 10^{-3}\ \text{m}^3}{1.38 \times 10^{-23}\ \text{J K}^{-1} \times (10 + 273)\ \text{K}}$ [1]

$= 3.1 \times 10^{23}$ molecules [1]

ii $\dfrac{p_2V_2}{T_2} = \dfrac{p_1V_1}{T_1}$ [1] but $V_2 = V_1$ so $\dfrac{p_2}{T_2} = \dfrac{p_1}{T_1}$

$\Rightarrow T_2 = \dfrac{T_1 \times p_2}{p_1}$

$= \dfrac{(10 + 273)\ \text{K} \times (0.6 \times 10^5 + 1.0 \times 10^5)\ \text{Pa}}{(1.1 \times 10^5 + 1.0 + 10^5)\ \text{Pa}}$ [1]

$= 216\ \text{K}\ (= -57\,°\text{C})$ [1]

6 a m, mass of each molecule; $\langle c^2 \rangle$, mean square speed of the molecules; T, Kelvin temperature of the gas [3]

b Each side represents energy – the average kinetic energy of a molecule. [1]

c $\frac{1}{2}m\langle c^2\rangle = \frac{3}{2}kT \Rightarrow \langle c^2\rangle = \frac{3kT}{m}$ [1]

$= \frac{3 \times 1.38 \times 10^{-23}\,\text{J K}^{-1} \times (-50 + 273)\,\text{K}}{5.4 \times 10^{-26}\,\text{kg}}$

$= 1.7 \times 10^5\,(\text{m s}^{-1})^2$

$\Rightarrow \sqrt{\langle c^2\rangle} = 410\,\text{m s}^{-1}$ [1]

Section 5 Nuclear decay

1 C

2 A

3 a Max 4 of: use of a GM tube and counter; measure background count with source not there; put source close to detector; put sheet of paper between source and counter (or move source away 3–7 cm); if this reduces count rate to background then source emits only alphas.

b Alphas go only 5 cm in air [1] OR the casing absorbs alphas [1].

4 a Nucleus has one less neutron OR one more proton [1].

b i First calculate λ:

$\lambda = \frac{\ln 2}{t_{1/2}} = \frac{\ln 2}{5730\text{ years}} = 1.21 \times 10^{-4}\,\text{years}^{-1}$ [1]

$N = N_0 e^{-\lambda t} \Rightarrow \frac{N}{N_0} = e^{-\lambda t} \Rightarrow \ln\left(\frac{N}{N_0}\right) = -\lambda t$ [1]

$\Rightarrow t = \ln\left(\frac{N}{N_0}\right) \div \lambda = \ln\left(\frac{2.3 \times 10^{-11}\%}{1.0 \times 10^{-10}\%}\right)$

$\div\ 1.21 \times 10^{-4}\,\text{years}^{-1} = 1.2 \times 10^4\,\text{years}$ [1]

ii Percentages of ^{14}C very small, so hard/inaccurate to measure OR proportion of ^{14}C in atmosphere might have been different then [1].

iii Half-life of ^{210}Pb closer to age of recent bones so significant proportion of ^{210}Pb will have decayed [1].

Section 6 Oscillations

1 B

2 D

3 a This is resonance [1]. For the amplitude to increase requires the frequency of the person's footfalls to match the natural frequency of oscillation of the bridge [1].

b i There is an external driving force caused by eddies in the wind flowing past the bridge [1] (called 'vortex shedding').

ii Max acceleration $= A\omega^2 = A(2\pi f)^2$

$= 0.90\,\text{m} \times \left(2\pi \times \frac{38}{60\,\text{s}}\right)^2 = 14.3\,\text{m s}^{-2}$ [1]

iii When max acceleration downwards exceeds g ($9.8\,\text{m s}^{-2}$) then the road surface is accelerating downwards faster than the car accelerates when falling freely; therefore it will lose contact with the surface [1].

Using $a = -\omega^2 x$ with $a = 9.8\,\text{m s}^{-2}$ and

$\omega = \left(2\pi \times \frac{38}{60\,\text{s}}\right) = 3.98\,\text{rad s}^{-1}$

$x = \frac{9.8\,\text{m s}^{-2}}{(3.98\,\text{rad s}^{-1})^2} = 0.62\,\text{m}$ [1];

so the car will leave the surface 0.62 m above the mid point of the oscillations.

Section 7 Astrophysics and cosmology

1 B

2 C

3 A

4 a We need an important little bit of maths here:

$P = kT^n. \Rightarrow \log\left(\frac{P}{\text{W}}\right) = \log\left(\frac{k}{\text{W K}^{-4}}\right) + n \times \log\left(\frac{T}{\text{K}}\right)$

Hence a graph of $\log\left(\frac{P}{\text{W}}\right)$ against $\log\left(\frac{T}{\text{K}}\right)$ should give a straight line with gradient n [1].

$\Rightarrow n = \frac{1.4 - 0.6}{3.1 - 2.9} = 4.0$ [1]

ResultsPlus
Examiner tip

If you're not very confident about maths, at least learn the equation above (don't worry about the units). A similar line appears in some radioactive decay questions.

b The electrical energy entering the lamp leaves it mainly through radiation from the surface. If it behaves as a black body we would expect it to radiate at a rate proportional to T^4 [1] according to the Stefan-Boltzmann law. The experiment confirms this, so the student is justified [1].

5 a Measurement of the distance to a distant galaxy is difficult, and involves considerable uncertainty [1].

b The age of the universe is approximately $\frac{d}{v} = \frac{1}{H_0}$ [1]

c The fate of the universe depends on its density/total mass [1], which due to gravitational forces is presumed to be slowing the expansion down. This means the Hubble constant would be changing over time [1]. Knowledge of how the Hubble constant has changed and will change in the future would allow us to calculate the fate of the universe [1]. (The expansion may also be changing due to the mysterious dark energy.)

Answers to the Unit 4 practice unit test

1	B	**2**	C	**3**	A	**4**	B	**5**	D
6	A	**7**	C	**8**	B	**9**	C	**10**	D

Section B

11 Max 5 of:

- inserting the magnet increases flux linked with solenoid
- rate of change of flux induces an emf (Faraday's law), hence current flows
- fast insertion ⇒ greater rate of change of flux linkage ⇒ larger emf and current
- removing magnet reduces flux linked with solenoid
- induced emf (and current) in opposite direction
- induced emf opposes flux linkage changes (Lenz's law)

12 a i Use of $Q = CV$ [1] 2.8×10^{-3} C

ii Use of $W = \frac{1}{2}CV^2$ [1] 8.5×10^{-3} J [1]

b Max 3 of:

- Charge capacitor through a large resistor at a constant rate,
- by using a variable resistor to gradually reduce the resistance in circuit
- monitor current with micro-ammeter, keeping current constant,
- record p.d. with voltmeter across capacitor,
- p.d. should increase at a constant rate.

13 Max 5 of:

- 'Plum pudding' model would predict all αs undergoing only small deflections,
- because in the 'plum pudding model' positive charge is uniformly spread throughout atom,
- αs rebounding indicated an interaction with a concentration of positive charge,
- Rutherford proposed the 'nuclear model',
- in this model the atom consists of a positively charged nucleus,
- the nucleus is tiny compared with the size of an atom.

14 a Use of $s = ut + \frac{1}{2}at^2$ with vertical motion [1] $t = 0.49$ s [1] Use of $s = ut + at^2$ with horiontal motion [1] $u = 1.9$ m s^{-1} [1]

b momentum conservation [1] correct substitutions [1] $m_b : m_m = 5.3$ [1]

c In an elastic collision E_K is conserved [1] calculation of E_K before and after [1], $E_{Kf} = E_{Ki} = 1.8 \times m_b$ [1]

15 a i A component of the push from the track acts towards the centre of the circle [1]

ii Equation for vertical equilibrium [1] Newton's 2nd law applied to horizontal forces [1] equation for tanθ [1] $v = 31$ m s^{-1} [1]

b Use of $F = \frac{mv^2}{r}$ [1] centripetal force = 3270 N [1] equilibrium of forces [1] 1120 N [1].

16 a The marble moves in a straight line [1].

b i Ball bearing is deflected into a curved path [1], as it is pulled towards the magnet [1].

ii Ball bearing is deflected into a less curved path, since it has greater speed [1] Same force acting for a shorter time produces a smaller change in momentum [1].

c Max 3 of:

- Charged particles are deflected in magnetic fields, similarly for ball bearings.
- Uncharged particles travel in a straight line, similarly for ball bearings.
- Path curvature depends upon particle momentum, similarly for ball bearings.
- Ionisation reduces particle energy ⇒ track length related to particle energy / moving sand away reduces ball energy ⇒ track length related to ball velocity.
- Only charged particles produce a detectable path, but marble and ball bearing produce a track in the sand [1].

17 a i Use of $Q = CV$ [1] 5.0×10^{-6} C [1]

ii Use of $E = \frac{Q}{4\pi\varepsilon r_0}$ [1] 3.0×10^5 V m^{-1} [1]

iii Radial lines from centre of dome [1]

b i Use of $E = \frac{V}{d}$ [1] $V = 3.75 \times 10^9$ V [1], $Q = CV$ to obtain 45 C [1]

ii Area of cloud surface is a tiny fraction of the Earth's surface.

18 a isotopes: same number of protons, different numbers of neutrons in nucleus [1]

b Conversion of mass to kg ($m = 3.3 \times 10^{-26}$ kg) [1] Gain in E_K = work done by electric field [1] $v = 3.1 \times 10^5$ m s^{-1} [1]

c i Use of $E = \frac{V}{d}$ to obtain $E = 5600$ V m^{-1} [1] Electric force equated to magnetic force [1] $B = 1.8 \times 10^{-2}$ T [1]

ii For plates drawn horizontal:
Electric field: vertical parallel equally spaced lines away from positive plate [1]
Magnetic field: acting perpendicularly out of paper if top plate is positive [1].

d i Magnetic force on electrons acts perpendicularly to plane containing B and v [1]. Force always at right angles to velocity, producing a centripetal acceleration [1].

ii $r = \frac{p}{BQ}$ [1] 0.21 m [1]

iii For magnetic field drawn acting into paper and ion beam entering horizontally from the left the beam is deflected downwards in arc of a circle [1]
Neon–22 ion beam has a larger radius of curvature since $r = \frac{p}{BQ}$ [1]

e Electron has mass – it is a particle (Thomson).
Electron has a wavelength – it is a wave (his son)
Electrons are best described by particle-wave duality (the current model) [Max 2]
Models change over time / are only as good as the evidence available [1]

Answers to the Unit 5 practice unit test

Section A

1	D	**2**	B	**3**	B	**4**	B
5	B	**6**	D	**7**	C	**8**	D

Section B

9 mass of air = density × volume = 1.3 kg m^{-3} × 0.20 m^3 = 0.26 kg [1]
$\Delta E = mc\Delta\theta$ [1] = 0.26 kg × 610 J kg^{-1} K^{-1} × (22 – – 18) °C [1] = 6300 J [1]

10 a $\frac{mv^2}{r} = \frac{GMm}{r^2}$ [1] and $v = \frac{2\pi r}{T}$ [1]
Cancel out m; then substitute for v and you arrive at $T^2 = \frac{4\pi^2 r^3}{GM}$ [1]

ResultsPlus
Watch out!
When you are deriving a formula like this question requires, you must make very clear the equations from which you are starting. They must be recognisable statements of physics in equation form. The steps you take from the starting equations to the desired end point must also be very clear.

b i Sub twice: $T_{HD}^2 = \frac{4\pi^2 r^3}{GM_{HD}}$ and $T_S^2 = \frac{4\pi^2 r^3}{GM_S}$ [1]
Careful rearrangement and cancelling $4\pi^2$ and G
$\Rightarrow \frac{M_{HD}}{M_S} = \frac{r_{HD}^3 T_S^2}{r_S^3 T_{HD}^2}$ [1] $= \frac{(3 \times r_S)^3 T_S^2}{r_S^3 (6 \times T_S)^2} = \frac{27}{36}$
= 0.75 [1]

ii The common centre of mass is much closer to the star than the planet, so the radius and circle of the star's motion are much smaller [1]; the time of orbit is the same for both; so the star orbits at a much slower speed [1].

c A larger planet means the common centre of mass will be further away from the star [1]; so the star will move round its circle faster [1]; so the Doppler shift will be great and so more detectable [1].

11 a activity $A = \frac{\text{power required}}{\text{energy per alpha}} = \frac{55\ \text{J s}^{-1}}{7.65 \times 10^{-13}\ \text{J}\ \alpha^{-1}}$ [1]
= 7.2 × 10^{13} α s^{-1} (Bq) [1]

b $\lambda = \frac{\ln 2}{t_{1/2}}$ [1] $= \frac{\ln 2}{1620 \times 3.15 \times 10^7\ \text{s}} = 1.36 \times 10^{-11}\ \text{s}^{-1}$ [1]

c $A = -\lambda N \Rightarrow N = \frac{A}{(-)\lambda}$ [1] $= \frac{7.2 \times 10^{13}\ \text{Bq}}{1.36 \times 10^{11}\ \text{s}^{-1}}$
= 5.3 × 10^{24} nuclei [1]

d mass needed
= 5.3 × 10^{24} nuclei × $\frac{226\ \text{g}}{6.02 \times 10^{23}\ \text{nuclei}}$ [1]
= 2000 g [1]

e Daughter nuclei may also be radioactive and emit energetic radiations OR may recoil passing on thermal energy [1].

12 a i If mass displaced distance x, (restoring) force $F = -kx$ [1]
and $F = ma$ hence $a = -\frac{kx}{m}$ [1]

ii Compare $a = -\frac{kx}{m}$ and $a = -\omega^2 x \Rightarrow \omega^2 = \frac{k}{m}$ [1]
We know $\omega = \frac{2\pi}{T} \Rightarrow T = 2\pi\sqrt{\frac{m}{k}}$ [1]

b i Resonance [1]

ii The forcing frequency matches the natural frequency of the oscillator [1].

iii $f = \frac{c}{\lambda}$ [1] $= \frac{3 \times 10^8\ \text{m s}^{-1}}{3.3 \times 10^{-6}\ \text{m}} = 9.1 \times 10^{13}$ Hz [1]

iv $T = 2\pi\sqrt{\frac{m}{k}} \rightarrow k = \frac{4\pi^2 m}{T^2}$ [1]
$= 4\pi^2 m f^2 = 4\pi^2 \times 1.67 \times 10^{-27}$ kg × (9.1 × 10^{13} Hz)^2 [1] = 550 N m^{-1} [1]

13 a $P = \frac{\Delta E}{t}$ [1] $= \frac{1.63 \times 10^5\ \text{J}}{347\ \text{s}}$ = 470 W [1]

b i $\Delta E = mc\Delta\theta$ [1] = 0.44 kg × 3800 J kg^{-1} K^{-1} × (96 – 12)°C = 1.4 × 10^5 J [1]

ii $P = \frac{\Delta E}{t} \Rightarrow t = \frac{\Delta E}{P}$ [1] $= \frac{1.4 \times 10^5\ \text{J}}{470\ \text{W}}$ = 300 s [1]

c i As temperature rises heat is lost from saucepan side and milk surface so less than 470 W goes into heating milk [1].

ii Measurement he made with the water also lost heat at similar rate, so the error due to heat losses in first measurement exactly compensated for heat loss in second measurement [1].

14 a $^{17}_{8}\text{O} + ^{1}_{1}\text{H} \rightarrow ^{14}_{7}\text{N} + ^{4}_{2}\text{He}$ [3]

b A white dwarf is a dead star, in which there is no longer fusion occurring (although it is still hot enough to be emitting significant radiation) [1]. There is no longer any hydrogen to fuse – but there are carbon, nitrogen and oxygen [1]. So the CNO process can occur, but only when hydrogen drawn from the neighbouring star arrives at the surface of the white dwarf [1].

c i mean $E_k = \frac{3}{2}kT$ [1] $= \frac{3}{2} \times 1.38 \times 10^{-23}\ \text{J K}^{-1} \times 10^7$
= 2.07 × 10^{-16} J [1]
$= \frac{2.07 \times 10^{-16}\ \text{J}}{1.6 \times 10^{-19}\ \text{J eV}^{-1}}$ = 1.3 × 10^3 eV = 1.3 keV [1]

ResultsPlus
Watch out!
You have to be eagle-eyed when reading questions – and spot when an examiner has slipped in an unusual unit – in this case not just eV but *keV*.

ii Gravitational force does work on the arriving hydrogen [1], which increases the internal energy of the gas [1].

d A standard candle is a cosmological event which has a fixed and predictable luminosity [1]. This implies that all novae of this type contain a similar quantity of hydrogen [1], are of similar size, and fuse at similar temperatures and rates [1].

15 a period = 3.0 s; maximum acceleration = 1.2 m s^{-2} [1]

b maximum acceleration = $A\omega^2 \Rightarrow$

$$A = \frac{\text{maximum acceleration}}{\omega^2} \text{ [1]}$$

$$= \frac{\text{maximum acceleration} \times T^2}{4\pi^2}$$

$$= \frac{1.2\,\text{m s}^{-2} \times (3.0\,\text{s})^2}{4\pi^2} \text{ [1]} = 0.27\,\text{m [1]}$$

c Graph should show:
cosine curve with initial positive value
displacement scale labelled [1]
maximum displacement shown as answer to **b**
timescale labelled [1]
two cycles shown, with period shown as 3.0 s [1]

16 a Two light nuclei combine to form a heavier nucleus [1]. Energy is released [1].

b High temperature and high density are both needed [1].
High temperature to overcome electrostatic repulsion between nuclei OR high density to maintain a high rate of collisions [1].

c There is a loss of mass [1]. This is converted to energy according to $\Delta E = c^2 \Delta m$ [1].

Index